SU/3081
15/3/81
AF841601
10 368 194 01
...technic Library
UNIVERSITY OF NORTHUMBRIA
AT NEWCASTLE LIBRARY
WITHDRAWN

6

THE SOLAR-HYDROGEN ENERGY ECONOMY

Beyond the Age of Fire

THE SOLAR-HYDROGEN ENERGY ECONOMY

Beyond the Age of Fire

Luther W. Skelton

VAN NOSTRAND REINHOLD COMPANY
NEW YORK CINCINNATI TORONTO LONDON MELBOURNE

NEWCASTLE UPON TYNE
POLYTECHNIC LIBRARY

620.93

Copyright © 1984 by Van Nostrand Reinhold Company Inc.

Library of Congress Catalog Card Number: 83-6664
ISBN: 0-442-28221-4

All rights reserved. No part of this work covered by the copyright hereon may be reproduced or used in any form or by any means—graphic, electronic, or mechanical, including photocopying, recording, taping, or information storage and retrieval systems—without permission of the publisher.

Manufactured in the United States of America

Published by Van Nostrand Reinhold Company Inc.
135 West 50th Street
New York, N.Y. 10020

Van Nostrand Reinhold Company Limited
Molly Millars Lane
Wokingham, Berkshire RG11 2PY, England

Van Nostrand Reinhold
480 Latrobe Street
Melbourne, Victoria 3000, Australia

Macmillan of Canada
Division of Gage Publishing Limited
164 Commander Boulevard
Agincourt, Ontario MIS 3C7, Canada

15 14 13 12 11 10 9 8 7 6 5 4 3 2 1

Library of Congress Cataloging in Publication Data

Skelton, Luther W.
The solar-hydrogen energy economy.

Includes bibliographical references and index.
1. Solar energy. 2. Hydrogen as fuel. 3. Energy development. I. Title.
TJ810.S4854 1984 33.79′23 83–6664
ISBN 0-442-28221-4

for Amy, Adam, Benjamin and Savanna

PREFACE

The Solar-Hydrogen Energy Economy: Beyond the Age of Fire is a departure from the usual approach to the presentation of energy technology in that it seeks to explain a number of high technologies in patterned relationships. These relationships define a different sort of energy economy than we have at present. Several assumptions, central to this work, not only motivated the effort, they also support many of the conclusions.

The first of these assumptions is that human beings can collectively make rational and effective decisions about big issues. While "satisficing" and "muddling through" may well be the common mode of decision making for most of us, about most issues most of the time, when the issue is important enough and well-enough articulated, the intelligence of the human spirit can shape the future.

A second assumption is that the technologies that we create structure our concepts of reality as we go about our daily lives. We are what we build in a very real sense. A third and related assumption is that the realities we shape with our technologies can differ greatly. Technologies that tend to restrict human growth and freedom through their centralizing effects are sometimes called "hard" technologies as opposed to those "soft" technologies that create freedom. This dichotomy is only valuable up to a point. Certainly some technologies are "harder" than others, but most often technologies are mixed in their effects. A transportation system like the automobile can liberate us from the isolation common to our forebearers, but it can also restrict our choices of where we can live and work by its pervasive effects on urban and suburban development.

The final major assumption is that a particular technology or system of technology is not justified just because it exists or because of the amount of resources that are invested in its development. Technologies can only be justified by their effects on people and on the environment. Choosing technologies is the great and serious responsibility of humankind.

On the basis of these assumptions, this book examines a range of technologies that together comprise a new energy economy. Called the solar-hydrogen energy economy, in its mature stage it does not depend on combustion or nuclear power. It is truly beyond the age of fire.

I am grateful for the research of the many scientists throughout the world whose professional lives are dedicated to the study of aspects of this system. Their exciting work continues to be one of the more encouraging signs that a positive and humane energy future is possible. I am particularly indebted to the visions of scientists like J. O'M. Bockris, Roger Billings, and Amory Lovins who continue to articulate the potential of these technologies.

Finally, I am deeply appreciative of the support I have received from my family and friends. Without that support I could never have done this work.

LUTHER W. SKELTON
Springfield, Illinois

CONTENTS

Preface vii

1. INTRODUCTION 1

2. CONCEPTS OF MATTER AND ENERGY 17

Light 17
Electricity 20
Hydrogen 27
The Solar-Hydrogen Connection 31

3. SUNLIGHT TO ENERGY 39

Photoelectrical Conversions 40
Biological Conversion of Sunlight to Energy 54
Mechanical Conversions 54
  Wind Conversions 55
  Water Power 61
Thermal Conversions 68

4. ENERGY TO HYDROGEN 76

Biochemical Conversions 76
  Coal 76
  Photosynthesis and Photosynthesis Mimicking 83
Thermal Conversions 86
Electrochemical Conversions 88

5. HYDROGEN TO ENERGY 100

Electrochemical Conversion—The Fuel Cell 100
  Applications: Electrochemical Conversion Devices 106
  Applications: Transportation 110
  Advantages of Electrochemical Conversion 112

Combustion Conversion of Hydrogen to Energy 115
How Hydrogen Burns 116
Safety Considerations 117
Domestic Residential Uses 119
Transportation Uses—Automobiles 121
Transportation Uses—Aircraft 129

6. STORAGE AND TRANSPORTATION OF HYDROGEN 136
Pressure Storage 139
Hydride Storage 143
Liquid Storage 147
Transportation of Hydrogen 149

7. THE OUTLINE OF A MODEL 155
Energy Use and Supply: 1985–2080 158
The Solar-Hydrogen System: Outline of a Model 160
Staging Primary Implementation, The T to T + 15 Year Period 166
Staging the Middle Period, T + 15 to T + 45 Years 168
Staging A Mature Solar-Hydrogen Economy, T + 45 to T + 95 Years 169

8. TECHNOLOGY DEVELOPMENT 173
Broad Requirements for Developing a Solar-Hydrogen Economy 176
Strategies for Technology Development 176
Planning 176
Compatability 178
Safety 179
Cost 180
Information 182
Basic Technology Delivery Systems 183
The Regional, Coal-Gasification, Hydrogen-Production and Distribution System 185
The Commercial Aircraft Network Using Liquid Hydrogen 186
The Modular Fuel Cell 187
The Hydrogen Electric Rail Transit System 187
The Solar Energy System for Use in Public Utilities 189
The Modularized Solar Hydrogen Residential Unit 190
Choosing Futures 191

Index 195

1. INTRODUCTION

For most of us, energy refers to a mysterious commodity we use to keep warm in the winter and cool in the summer and to transport us from place to place. Usually our awareness does not extend beyond these clear and vital perceptions of satisfying simple, immediate demands. When energy needs become expensive, concern is most often focused on the local distributor of gasoline or the regional public utility. Few persons realize the extent that their comfort, convenience, and ultimately survival are part of a global pattern of energy production, storage, distribution, and use. There is little public understanding of this global energy system, and yet there is virtually no way to live outside its influence. The global energy system which now meets most of our needs can be termed a *carbon-combustion energy economy.* This term indicates both an economic common denominator and a technological process.

An energy economy is a series of source to end-use energy patterns defined by particular processes. It involves several distinct and identifiable aspects including source, conversion, energy storage, and methods of delivery.

An energy source is a quantifiable grouping of matter or pure energy which can be used to satisfy immediate human needs or which can be converted and transported for future use. There are five primary energy sources which fit this definition: biological sources, geological sources including fissionable elements, sunlight, wind and water. It can be argued of course that all of these are really aspects of solar energy. Nonetheless, from both a political and an economic perspective they are treated as distinct sources.

Most energy sources are not usable in their naturally occurring state. They must be converted into a form which can either be used or stored and transported to a point of use. Several conversion processes are presently applicable: chemical, photoelectrical, electromechanical, mechanical, and nuclear fission and fussion. These various methods of conversion are applicable to particular energy sources and many of them will be discussed in future chapters.

The form of energy which results from conversion may be viewed as an energy medium a concept which may ecompass both storage and application. Three of these energy mediums are of particular interest to us: electricity, heat, and motion.

Finally, energy must be delivered to the site of use. It may be delivered in its source form or its converted form. There are four major methods of energy delivery in common use: physical delivery via mechanical conveyance (e.g., ship, truck, rail, etc.), pipeline, electrical transmission line, and radiation.

Once an energy economy is established it defines by its nature a particular package of goods and services. These goods and services are distributed to the whole society. Goods and service packages are quantifiable and range from what could be called the primary package containing a variety of goods and services to packages containing only one or two. The primary energy packages in today's society provide light, heat, motion and electricity.

The carbon-combustion energy economy consists of two major forms which differ in their respective energy processes from each other beginning with the conversion process. The most basic and the oldest of these forms uses combustion as a conversion process to create heat and light which is used onsite or is transported in pipes as hot water or steam. This form of the economy provides lighting, heating, cooking, transportation, and energy for industry; however, technologies of delivery are limited either to delivering the carbon compound itself (wood, coal, gas or oil) to the use site or to delivering heat as steam or hot air to the site. In the case of a major energy processor system such as the petroleum industry, oil is mined and piped for multiple uses, most common for conversion to gasoline. Petroleum in various forms is delivered to use sites by pipeline or physically via mechanical conveyance.

The second form of the carbon combustion economy involves major secondary conversions. In this case the biomass (oil, coal, or natural gas) is converted to heat through combustion, and the heat is converted into mechanical energy through gas expansion (steam).

Mechanical energy is then converted into electricity through electromagnetic force field manipulation in generators. The electricity is then delivered to the use site by electrical transmission lines. This form of the carbon combustion energy economy delivers a primary service package with great convenience to the user.

These two forms of the carbon combustion energy economy differ in their use of the heat and light produced. In the former case the heat or light of combustion is used directly to satisfy an energy need for warmth, lighting, transportation, or industry; in the latter case the heat is converted to electricity. Both forms however depend upon burning carbon compounds.

Carbon is a chemical element common to all life processes. It is present in all plants and animals. In its various compounds carbon has been used to provide clothing and shelter throughout human history. The rapid oxidation, or combustion, of carbon in the form of wood was the foundation of our present energy economy. Rather recently in the human domination of the planet it was discovered that the carbon and hydrogen compounds, called hydrocarbons, would also burn. These compounds (coal, oil, and natural gas) are the remains of long dead trees and animals compressed into the earth by the movement of glaciers during previous ice ages.

Throughout human history, carbon has been the currency of the energy economy and combustion has been the process used for converting it into usable energy. So available was carbon in the form of trees that for a long period wood was used as the only fuel. However, as the human population increased, the forests were cut until much of the easily accessible wood was consumed. Readily available coal was then used to provide heat for the developing industrial world. As coal became scarce and its labor-intensive mining placed more strain on society, oil became the dominant carbon compound in the energy economy. But the carbon-combustion energy economy has certain characteristics which make it a poor system for global energy production and distribution. These characteristics are now becoming so pronounced in their effect on both society and the biosphere that we must begin to consider developing a new energy economy.

There are two problems associated with the carbon combustion energy economy: pollution and scarcity. When coal and oil are burned several chemical compounds are produced. Some of these compounds, such as sulfur dioxide (SO_2) which is a primary cause of acid rain, are dangerous to human beings and to other life systems as well. Ironically, it is the production of carbon dioxide (CO_2), an essential part of the life process, which may be the most dangerous. Carbon dioxide is used by green plants in their life process; however, when it is produced in excess it migrates to the upper atmosphere and acts as a trap for solar heat. This creates the greenhouse effect. While hard proof is not yet available, some scientists are convinced that the greenhouse effect may be the greatest environmental threat ever faced by the planet.[1]

Other dangerous pollutants produced by hydrocarbon combustion are the oxides of nitrogen. These oxides chemically deplete the ozone layer of the biosphere. Since the ozone in the upper atmosphere is the primary shield protecting the planet from a potentially lethal amount of radiation, the deterioration of this ozone layer must be viewed as very dangerous.

A second major problem inherent in the carbon-combustion economy is the developing scarcity of hydrocarbon fuel. This problem of scarcity is a function of the costs associated with the mining, pumping, and transporting of oil.

Oil of high quality is still readily available in North Africa and the Middle East and there may be other parts of the world where large quantities might be located, but we are becoming painfully aware of a national dependence on oil produced in foreign countries. This dependency requires us to risk potential military conflict every time the reliability of our oil supplies is threatened. Since much of this oil is produced in countries which have a legacy of political instability, our national security is proportionally destabilized.

There is apparently adequate coal available to sustain the carbon-combustion economy in the United States for well over a century. However, mining and transporting this coal are becoming much more expensive. One reason for this increasing expense is that much of the readily available coal lies under prime farm land; in order to mine it cheaply, this land must be stripped. There is no way to replace or rebuild prime farm land, so intricate and balanced is its structure. The cost of the destruction of prime farm land must be counted as an expense in strip mining coal.

Finally, and in the long run, hydrocarbons have many more valuable uses. Plastics, synthetic fabrics, and the full range of products which are necessities both now and in the future are all made from hydrocarbons. By using hydrocarbons to produce heat we are irreversibly consuming materials which are important to our survival.

The dangers inherent in the carbon-combustion economy grow greater every day, yet we have the capability of developing different energy economies. This book is about one such energy economy which seems to offer a great range of human options with little danger to human life and environmental quality. It is an energy economy which in its mature stages can take us beyond the age of fire.

The concept of a solar-hydrogen economy in one form or another has fascinated scientists for over a century. In recent years, as a result of the work of J. O'M. Bockris, E. W. Escher, and many others, the concept has been given clearer form. Thousands of scientists and engineers throughout the world are now working to develop various aspects of a solar-hydrogen energy economy.

The concept of a solar-hydrogen energy economy itself is deceptively simple. In a mature solar-hydrogen economy, solar energy and hydrogen are used as the energy common denominators supplanting carbon. While hydrogen can be produced using any energy source, solar energy is used as the energy source in a mature economy. The system creates virtually no environmental pollution, is based on renewable resources, and is almost totally independent of national boundaries. However, beneath this simplicity, there is both technical and economic complexity.

In a solar-hydrogen economy the energy of the sun must first be converted into usable energy which can be used onsite, stored for future use, or transported for use to some distant place. There are several ways of converting sunlight directly into usable energy. The most obvious way is to convert it to heat. The range of technologies which involve the conversion of sunlight to heat are often referred to as *solar thermal* processes. These technologies range from the familiar greenhouse to extremely powerful solar furnaces which reach high temperatures. However, heat energy is difficult to store and transport. Due to the transportation and storage difficulties, solar thermal devices have been limited primarily to onsite, energy-use systems or high-temperature conversion systems which make steam for electrical generators.

Solar energy can also be converted into electricity in a number of other ways. The motion of air for windpowered generators, the motion of water for hydropowered generators, and the temperature differentials of the ocean are all aspects of solar energy, and all of them can be used to make electricity. In addition to these possibilities, sunlight can also be converted directly into electricity through the photoelectric effect. *Photovoltaics* is a term often used for the process of electrical current generation based on the photoelectric effect.

The conversion of light into electrical current has been known to be possible since the 1800s, but it was the space program in the 1960s and 1970s which encouraged the development of this technology. The process has been closely tied to the electronics industry because the most common material used in the conversion has been pure, single crystal silicon which is also used to make transistors and computer microchips. Very small amounts of pure single crystal silicon are needed to make electronic circuits, and a relatively high price for the material can be easily tolerated by the electronics industry. Much larger amounts however are required to make photovoltaic devices which produce a significant electrical current. The price of photovoltaic produced electricity has remained prohibitively high for residential and industrial purposes, in part because most of the single crystal silicon used in photovoltaic devices is produced using the costly technology necessary for the high purity silicon used by the electronics industry.

The situation has begun to change in recent years. Considerable research has been done in the area of photovoltaics in the mid- to late-1970s, and major breakthroughs are being made which should reduce the price of photovoltaic electricity enough to make it competitive in certain residential and industrial markets. The U. S. Department of Energy established a Low-Cost Silicon Array (LCSA) project in the 1970s. This project was designed to encourage the development of cost-effective photovoltaic devices. The dollar cost of photovoltaic devices is usually measured in terms of the amount of power (watts) which can be generated at peak sunlight. In 1976 the price of photovoltaic power was about $20 per peak watt; by 1978 it had dropped to about $10. The goal set by the LCSA is $.20 per peak watt by 1990. Recent developments by Texas Instruments, RCA laboratories, researchers at Texas A & M University, and others make that goal appear somewhat conservative.[2] A competitive and effective photovoltaic system is an essential part of a solar-hydrogen economy because such a system could create a dependable and cheap source of electricity.

Electricity which is generated by solar energy has the advantage of being usable in both centralized and decentralized energy systems. In decentralized systems individual homes can be equipped with rooftop photovoltaic collectors which convert sunlight into electricity to supply all of the residence's needs for electricity. Electricity can be stored as hydrogen for future use in order that the system can be effective even during long periods of cloudy weather. In 1979 *Solar Age* magazine indicated that such a system might reach efficiency and production cost levels that would allow production of electricity for as little as $.50 a peak watt by 1986.[3] Developments at RCA laboratories, Texas Instrument Corporation and Texas A & M University increases the likelihood of reaching the goal of $.50 per peak watt considerably sooner.

Photovoltaic systems can also be effective on a much larger scale in the centralized production of energy for use by industry and commerce. Large areas of unused desert regions can be converted to solar-electric farms. The electricity thus produced can be converted into hydrogen for transportation and storage for later use in distant areas.

There is no pollution in the production of electrical energy by photovoltaics except for that produced in the manufacture of cells. What sunlight is not used in the production of electrical current is reflected away or dissipated as heat. Photovoltaic systems have long use lives, no moving parts, and can be built in modules of virtually any size.

Electrical energy, no matter how it is produced, is difficult to store for future use. The primary method of electricity storage is as chemical energy in batteries, but this method has serious limitations in both efficiency and capacity. Most often electricity is not stored; it is produced, transported, and used immediately. When demands for electricity are uneven, as they almost always are, provisions must be made for increasing or decreasing production. The problem of providing varying amounts of electricity according to load demand is eased somewhat by transporting electricity from one demand area to another through the electrical grid system. However, transporting electricity has its own serious drawbacks. Electricity is moved from place to place in the electrical grid system by conduction through wires. This conduction of electrical current is expensive in energy terms. As an electrical current moves through a wire, energy is consumed in the movement. Some of the electricity is used as heat and some is used in the creation of a magnetic field around the conductor. It is estimated that for the small consumer of energy only about twenty-nine cents of every dollar spent on electrical energy actually buys energy. The remainder is expended on transportation and distribution (about $.43), equipment maintenance (about $.21), and profit (about $.06).*[4]

The problems of varying load demand and energy loss through transmission of electrical current have been with us so long that they have come to be accepted as inherent in any electrical energy system. However, part of the problem can be solved by using hydrogen as a means of storing and transporting electrical energy. Hydrogen has a number of properties which make it a very attractive means of energy storage and transmission; these properties will be discussed in more detail later. At this point it is important to indicate a particular relationship which exists between electrical energy and chemical energy, particularly the chemical energy inherent in hydrogen.

Hydrogen, in its most common form on earth, is found as part of the water molecule, H_2O. If an electrical current is passed through water, the hydrogen and oxygen will separate as free elements. They can then be collected and stored for future use. This process is called *electrolysis* of water and is commonly used to produce hydrogen and oxygen in many parts of the world, particularly in areas where electricity is produced by water power.

Hydrogen in a free state is relatively stable; however, it readily combines with oxygen to form water and release energy. In the process of water formation a considerable amount of energy is released. Pound for pound hydrogen produces about three times the heat en-

---

* The $.01 which appears to be missing is probably due to arithmetic variation resulting from differing growth rates of the components in the analysis. For a more complete description of this study see: Amory B. Lovins, *Soft Energy Paths, Toward A Durable Peace,* New York, Harper and Row, 1977.

ergy of petroleum when burned and can be used for most of the same purposes. There is an even more useful energy process of water formation however. The energy released in the process of making water will be in the form of electricity if electricity is used to start the process. If hydrogen and oxygen (or air) are placed together in an electrical field in water they will combine to make water and electricity. The device for creating this effect is called a *fuel cell.*

The use of hydrogen as a means of storing and transporting energy is a relatively new concept which began to attract attention in the 1930s. The many uses of hydrogen, and its flexibility as a fuel, have led some authors to suggest that hydrogen should become the dominant aspect of the economy, hence the term *hydrogen economy.** However, some of the public and many energy planners retain a bias against hydrogen which may result primarily from a misunderstanding of its nature. Part of this misunderstanding is related to the early uses of hydrogen. People are generally aware that hydrogen was used to provide lift for lighter-than-air craft in the 1930s and that one of these craft, the Hindenburg, crashed and burned while docking at Lakehurst, New Jersey, in 1937. This event was heavily covered by the media and that coverage may have distorted the public perception of the event to the extent that hydrogen has been viewed as the most dangerous of fuels. In addition, many people associate the use of hydrogen with the hydrogen bomb. This unfortunate association is absolutely in error. We will examine the public perception of hydrogen in a later section on technology development.

Hydrogen contains about three times the energy of gasoline for equal weight and it ignites at a much lower temperature. However, it is lighter than air and this lightness means that its combustion rapidly expands upward, whereas the combustion of gasoline fumes remains low to the ground. Hydrogen is potentially no more dangerous than

---

* J. O'M. Bokris is generally acknowledged to have popularized the term *hydrogen economy* in an article in *Chemical and Engineering News,* October, 1972. Previously in 1962, Bokris suggested a solar hydrogen system which would supply American cities from floating energy platforms. The idea of the potential use of hydrogen as a means of storing and transporting energy however probably dates back to the discovery of the element. Mention is made of this potential in Jules Verne's book *Twenty Thousand Leagues under the Sea.* Also see Derek P. Gregory, "The Hydrogen Economy," *Scientific American,* June 1973, Vol. 228, No. 1, pp. 13–21.

natural gas and it could be considered less dangerous than gasoline under most circumstances.

The concern about hydrogen on the part of energy planners is usually associated with hydrogen storage and transportation. These difficulties are all more or less related to the lightness and chemical activity of free hydrogen.*

Hydrogen is lighter and more chemically active than natural gas; however it can be transported and stored on a large scale in exactly the same manner, with some modification to accommodate its lightness and chemical activity. Much of the system now used to transport natural gas could be used to transport hydrogen.[5]

The technology for transporting and storing hydrogen can be improved considerably—we will consider possibilities in a later section. However, even using present technology, the use of hydrogen offers about half as costly a means of transporting energy as does electrical current.

Hydrogen may be kept indefinitely and used whenever and wherever there is need. There are many ways of storing hydrogen, some conventional and others verging on the frontier of science. The technology for the conventional storage of hydrogen is basically the familiar technology of natural gas. It can be stored in the ground or in tanks above ground. Larger scale storage could easily be accomplished in depleted oil and gas fields or in abandoned mines.[6]

Hydrogen, due to its unique chemical nature, easily combines with other elements for storage as *hydrides.* The most common hydride storage at present is as an alloy of iron and titanium which can store more molecules of hydrogen than are contained in an equal volume of liquified hydrogen. Hydrogen storage in *glass microballons* and other new technologies is being developed.[7]

Hydrogen can be transported through pipes similar to those used for natural gas, and many of the major natural gas pipelines in the United States are now adequate to handle hydrogen either in a pure form or mixed with natural gas. We noted earlier that a considerable

* Wilson Clark shared with many other energy planners a real concern about the problems of handling hydrogen. Such concern is understandable, and the storage and transportation of hydrogen is a major focus of research in the field of hydrogen energy. See Wilson Clark, *The Energy and Defense Project. Energy, Vulnerability and War* (Washington, D.C.: Federal Emergency Management Agency, 1982).

amount of electrical energy is used in moving electricity from place to place. Most of the energy which is transported through pipes as hydrogen is not lost in the process of moving from one place to another. If electrical energy is converted to hydrogen and then transported through pipes and converted to usable energy, some of the loss can be recovered. The amount recovered is of course a function of the distance the energy is transported and the cost of hydrogen production.

In the model discussed later in this book the conversion, storage, and transportation of solar energy and hydrogen are combined through a five-stage process to provide the technical foundation for a mature solar-hydrogen economy. Solar energy can be converted into electricity or heat through solar thermal technology, photovoltaics, water power, or windpower. In the transition stages coal is also converted to hydrogen. Where possible the energy thus produced is used at the site of production or transported short distances as electrical current. When onsite use is not possible, either because the energy is required at some distance from the site or because the use demands are intense (such as in industry) or widespread (as in transportation), then solar energy can be converted to hydrogen and stored for future use or transported to a distant site. Hydrogen can then be reconverted to usable electrical energy in fuel cells. These fuel cells can be used to power electric cars, buses, and trains; or they can supply the energy needs of industry, commerce, and residences. This system requires only sunlight and water and the resources to build the conversion devices.

This model is not a prediction of an energy future; it is presented rather as a way of examining technological alternatives which go beyond the carbon-combustion economy. Most people, including energy planners, are aware that oil and coal are both the results of fossilized plants and animals. However there is still a basic assumption that the supplies of these biological remains will last indefinitely, or that they will last at least for the rest of the life span of this generation and the next. The supplies of oil and coal will probably last forever, but only because their extraction from the earth and their subsequent use will become so expensive that human society will no longer be able to pay the price.

When the full costs of the use of fossil fuels are not accounted for in the market system they are paid for in other coin such as environmental destruction and a decrease in human health. These costs are often referred to as *market externalities,* some of which can be brought into the economic system or internalized.[8] For example, the cost of air pollution in the production of electricity through burning coal can be added to the price of energy and the increased revenue used to build scrubbers. Some costs however are beyond internalizing. These costs are irreversible consequences of policies or events and can never be paid for. A global nuclear war fought over oil in the Middle East would be one of these. There is no price which can pay for an earth made permanently uninhabitable by a nuclear war.[9]

Bokris holds that a transition from oil to a new energy source must be accomplished in the next 50 years as the result of depletion of natural gas and oil. He suggests that the necessary transition to any of the three major available alternatives—coal, breeder reactors, and solar will cost about $1000/kW for each unit of conversion machinery. He notes however that when market externalities are considered, the solar-hydrogen option becomes not only the most attractive, but also the least costly.[10]

Because there is currently a general acceptance of an energy economy based on fossil fuels which are considered limitless, we are forced to evaluate all alternatives to that economy in terms of the prices of that economy. The cost of an alternative energy economy such as the solar-hydrogen system are evaluated in terms of the projected cost of oil and coal, but these costs, which the industry projects, may not be realistic due to an inadequate accounting for market externalities.

In the model presented in this book we will examine one approach to a solar-hydrogen economy. This approach postulates the staging of an energy economy which in its mature form does not use combustion except for the burning of hydrogen in jet aircraft. In this mature solar-hydrogen energy economy the source of energy is sunlight which is converted to usable heat, electricity, or motion. These products of sunlight are then converted into hydrogen which is used to store and transport solar energy. The hydrogen is finally reconverted electrochemically into electricity for use in homes, industry, and transportation. The system involves no burning of fuel, and its primary product, in addition to usable energy, is water. As we have noted this system has the potential flexibility to be either independently dispersed or centralized to whatever degree desired. The technologies for such a system are available now, though much research and development is needed in order to refine the various systems and to make them economically feasible.

We will examine these technologies in four broad categories and then organize these technologies into six *solar-hydrogen technology subsystems* phased over a one-hundred year time period. The vehicle for this examination will be a descriptive plan for implementation.

Three of the four broad categories in which technologies are considered are energy conversion processes; the fourth is a category specifically involving the technologies for storing and transporting hydrogen. The technologies considered have been selected from a wide range of possible choices. The criteria for this selection are admittedly judgmental and are based on the author's perceptions of comparative technological simplicity, energy efficiency, research and development potential, environmental security, and economic and social possibility.

In order to understand better the technologies involved in the system we will first consider some principles of matter and energy upon which they are founded. We will review the ideas which led to our present-day understanding of light and electricity, and then examine hydrogen in some detail in order to understand its potential as an energy storage system. Finally, we will consider the solar-hydrogen connections which make this system possible.

Next we will examine the technologies involved in the conversion of sunlight to usable energy. Four major conversion processes are considered. The first process, the biological conversion of light into energy, is the photosynthetic production of living matter. We have selected coal as the only product of the process with applicability to the solar-hydrogen economy in the immediate future. In the long run a number of living materials have applicability to the staging of a solar-hydrogen economy. Coal gasification to hydrogen however has applicability as a transition technology during the early staging of the system.

The second process, mechanical conversions of sunlight into energy, involves wind power and water power. Five technologies are considered in this category: wind generation of electricity, hydrogeneration of electricity, wave generation of electricity, tidal generation of electricity, and ocean thermal technology.

The third process, thermal conversion, is the conversion of sunlight into heat. Two technologies are selected from this process. The first of these is the passive and active use of solar heat as a design feature in habitats. This technology is projected to reduce dependence on electrical energy. The second technology is the use of sunlight to produce intense heat in solar furnaces.

The fourth and final solar conversion process considered is the photoelectric process. The one technology selected in the process is the photovoltaic generation of electricity.

In chapter 3 we will consider processes for the conversion of energy to hydrogen. The first of these processes is biological and involves two technologies: coal gasification to hydrogen as a staging process, and photobiotics and photobiotic mimicking which have system application in the more distant future. Another conversion process is the thermal dissociation of water. Finally we will consider the electrochemical dissociation of water in what is commonly called the fuel cell.

The technologies discussed in chapters 2 and 3 are combined into seven primary conversion technology patterns leading to the production of hydrogen. In the order of their staging in the model these are:

1. Coal → coal gasification → hydrogen
2. Hydropower → electrochemical disassociation of water → hydrogen
3. Photovoltaic electricity → electrochemical dissociation of water → hydrogen
4. Wave and tidal generation of electricity → electrochemical dissociation of water → hydrogen
5. Ocean thermal generation of electricity → electrochemical dissociation of water → hydrogen
6. Solar thermal energy → thermal dissociation of water → hydrogen
7. Various bacteria → photobiotic conversion → hydrogen

Next we will examine two primary conversion technology patterns for converting hydrogen to energy, the *electrochemical process* (fuel cells) and the *chemical process* (combustion). These technologies are described and discussed with regard to their applicability.

We will then examine processes for storing and transporting hydrogen. This examination leads to the selection of three categories of hydrogen storage technologies: *pressure storage, liquid storage,* and *chemical storage.* Four basic technologies have been selected in order of the staging:

1. Tank and pipe pressure storage and liquid storage
2. Underground storage
3. Metal hydride storage
4. Microballon storage

One basic technology for transporting hydrogen is examined:

1. Transportation in pipe

After considering the technologies we will describe the general outline of a model plan for implementing a mature solar hydrogen energy economy over a period of several decades. Finally we will examine some of the problems and possibilities of implementing the technologies required for the six subsystems.

In total, this book is an effort to aid in the ongoing examination and explanation of the energy future from a positive point of view. The fundamental assumption is that there is an alternate energy future which can increase human choice, safety, and quality of life and which does not threaten the existence of the biosphere or consume valuable resources.

NOTES

1. *Fragile Greenhouse, The Economist,* March 1982.
2. Paul Danish, "Low Cost Silicon and PV Devices Could Result from SERI Process," *Solar Times,* December 1981, p. 1. Jeffrey L. Smith, "Photovoltaics," *Science,* 212, June 26, 1981, pp. 1472–1478. Carol L. Yaws et. al., "New Technologies for Solar Energy Silicon: Cost Analysis of VEC Siliane Process," *Solar Energy,* 22, pp. 547–553. The above articles are only a representation of the great number of optimistic reports concerning lower cost photovoltaic devices.
3. Lee Muller, "Hubbub at Texas Instruments," *Solar Age,* December 1979, pp. 16–17.

4. Amory Lovins, *Soft Energy Paths* (New York: Harper and Row, 1977) p. 88.
5. Gordon N. Patterson, *The Future Role of Hydrogen Fuel in an Electrical Society,* (U.S. Department of Commerce, October 1979).
6. Yaws, p. 13.
7. "Hydrogen Update, Microsphere Storage Investigated," *Hydrogen Progress Magazine,* Winter 1979–1980.
8. For an insightful discussion of the concept of market externality, see Stahrl Edmunds and John Letey, *Environmental Administration* (New York: McGraw-Hill, 1973).
9. Wilson Clark, *The Energy and Defense Project: Energy, Vulnerability and War* (Washington, D.C.: Emergency Management Agency, 1982). For a unique perspective on the relation of energy and national security.
10. J. O'M. Bockris, *Energy Options, Real Economics and the Solar-Hydrogen System* (New York: John Wiley & Sons, 1980), p. 408.

2. CONCEPTS OF MATTER AND ENERGY

A review of some basic concepts of matter and energy is a useful first step toward an understanding of the nature and potential of a solar-hydrogen economy. Of particular interest are the natures of light, electricity, and hydrogen.

LIGHT

Energy has many forms, and life systems have the ability to respond to these forms in different ways. Human beings are directly sensitive to a small part of the whole energy spectrum which composes existence. Our eyes have evolved a sensitivity to that part of the energy spectrum which ranges from red to violet light. Between these two extremes of perception lie all the colors of the visible spectrum.

We can think of all energy as wavelike motion which is a manifestation of the oscillation, or vibration, of atoms. This motion can be measured in terms of intensity and number of repetitions per unit of time, and we can visualize it as a series of waves in any medium. A tight string or wire when plucked vibrates in waves, as does water in a pond when disturbed by a thrown pebble. If we think of light as a series of waves in some imaginary medium, we can then express its energy in terms of intensity and repetition per unit of time. The two characteristics of a wave which define these qualities are *amplitude* and *frequency*.

The amplitude of a series of waves is the height of the waves, and when we think of light as waves this amplitude is called intensity, or brightness. The frequency of a series of waves is the number of waves (*cycles*) per unit of time (*seconds*). In terms of light, frequency defines color. We may visualize the length of a wave as the distance between the top of one wave to the top of the next. This is called *wavelength* and is another way of describing frequency; the longer the wavelength per given unit of time, the lower the frequency of the wave. Wavelength and frequency thus have an inverse relationship with each other. Visible light with the longest wavelength (lowest frequency) is red. Visible light with the shortest wavelength is violet. In between the violet and red extremes of visible light, colors range from blue to green, yellow, and orange, covering the whole spectrum as wavelength increases and frequency decreases.

Visible light however is only a small part of the energy spectrum. Wavelengths in the whole energy spectrum may range from 10^5 meters (radio waves) to less than 10^{-14} meters (gamma rays), with visible light ranging from about 10^{-6} to almost 10^{-7}meters (Figure 2.1).

The wave theory of light was thought to be a true representation of reality for many years, but the test of a theory is how well it can answer questions about the phenomena it is designed to explain. As observations of the various phenomena explained by wave theory yielded more knowledge, the theory worked less well.

Early indications that the wave theory of light was not explaining some things very well came at the turn of the century. In 1839, the French scientist Becquerel noted that when light fell on one side of a simple battery it caused a current to flow. In 1888 Heinrich Hertz observed that when ultraviolet light struck the zinc plate in a vacuum tube a negative charge was produced. Both of these observations showed a relationship between light and electrical activity. It was determined by Lenard in 1900 that the electrical charge which devel-

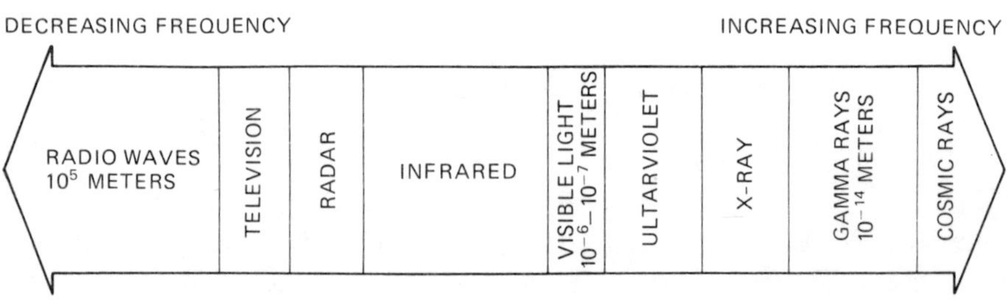

Figure 2.1 The electromagnetic energy spectrum

oped when light hit particular metals was caused by electrons in the metal which had been freed by the light. However, to the surprise of most scientists, each frequency of light (color) affected different metals. Some colors of light freed electrons from some metals but not others. While ultraviolet light dislodged electrons from copper, lower frequency light waves would not. The energy of these released electrons was a function of the wavelength of the light and of the nature of the metal, but not the intensity or brightness of the light. The maximum energy possessed by the most energetic electrons excited by a particular color could not be increased no matter how bright (intense) the light was made. It was also observed that the release of electrons was instantaneous when a metal was struck by the proper wavelength of light. The metal did not seem to need time to reach an energy level or to "heat up."

The wave theory model of atomic oscillation did not seem to fit these observations. According to the wave theory, the motion of the light wave should have set electrons in a metal's surface into oscillatory motion. The amount of energy imparted in this motion should have been proportional to an electrical field, produced by both intensity and frequency of the light wave. The brighter the light (intensity) the greater the amount of energy which should have been delivered to the electrons. In addition, since energy, according to the theory, was supposed to be cumulative and continuous, a dim light of any frequency should have imparted energy to a metal's electrons if it shone long enough; surely a bright light of whatever frequency should have imparted some energy. However, light did not act this way. No matter how intense, a red light would not free an electron from copper; and no matter how dim, an ultraviolet light would still free electrons from that same metal. As a result of these observations Max Planck, a twentieth-century scientist, postulated that:

Atomic oscillation exists only in discrete quantized energy states specified by the relationship:

$$\mathrm{En} = nhf, \qquad \mathrm{n} = 0,1,2$$

f = frequency of oscillation
h = a proportioned constant
En = the allowed energy state of oscillation

Based on Planck's postulate, Albert Einstein proposed in 1905 the theory of the photoelectric effect, for which he received the Nobel Prize. Einstein postulated that when an oscillator (an electron) moves from a higher energy state to a lower state, a discrete package of energy is emitted equal to Planck's constant (h) times the frequency (f). He called this *quantum* (package of energy) a *photon*. He suggested that a photon has the basic characteristics of a particle and that it maintains its localized character as it travels through space at the speed of light. When a photon collides with an electron in a metal it gives its entire energy to that electron. If that energy is sufficient the electron will be raised to a higher level, or released, with a *photoelectron* energy equal to the initial energy of the photon minus the *binding energy* of the electron. Einstein's observation, which he called the photoelectric effect, is the basic principal of *photovoltaics:* the generation of electricity directly from sunlight. We will consider the technology of photovoltaics in more detail in chapter 3.

Photons resemble particles, but they also conform to many of the principles of wave theory, particularly when scattered. Light must be viewed as both a particle and a wave, though neither in itself seems to represent reality.

If visible light resembles localized particles, then it makes sense that the rest of the energy continuum does also. However, it seems clear that quantum photons of energy are not charged pieces of matter which are physically carried along through a medium, whether substantial (like water) or theoretical (the *ether* which was earlier thought to pervade space). We cannot isolate a photon and study it as a particle. Light and perhaps all energy is observable only in its effects.

We have examined that part of the energy spectrum which we call light and have observed that it resembles both a wave and a particle. In order to begin to understand another familiar form of energy, electricity, we must understand some of the ways that energy and matter are related.

ELECTRICITY

In 1897 Sir J. J. Thomson postulated that atoms are composed of corpuscles distributed in a sphere with a positive charge. These corpuscles, or electrons, were held apart by mutual repulsion. This is

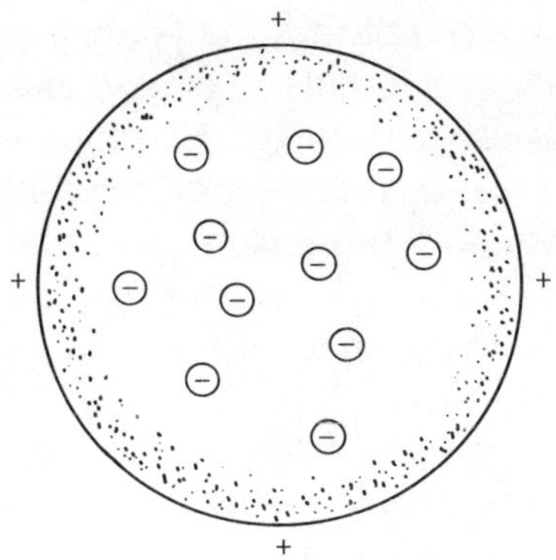

Figure 2.2 Raisin cake concept of the atom

called the *raisin cake* model of an atom (Figure 2.2). Thomson's model predicted that when an atom-sized particle collided with an atom it would always penetrate it completely.

In 1910 E. Rutherford bombarded a thin gold foil with alpha particles (nuclei of helium) and measured their deflection. One alpha particle in every 20,000 was deflected from the gold atoms without passing through them. This phenomenon destroyed the Thomson model because it showed that the atom had to contain something much more dense than electrons in order for the alpha particles to bounce back. It led Rutherford to postulate the existence of the nuclear atom in which the electrons moved in a circular orbit about a dense center, a miniature solar system (Figure 2.3). Alpha particles bounced back when they hit the dense center or nucleus of an atom. The theory however also held that atoms radiated energy in a continuous pattern which made up the energy spectrum. If the atom were really a miniature solar system with the electrons held in place by a

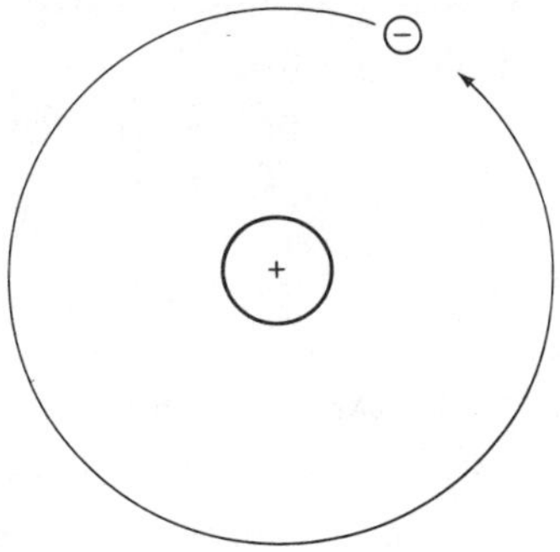

Figure 2.3 Nuclear concept of the atom

combination of centripetal force and charge attraction, then the position of electrons would depend upon a balance between these two forces. If however electrons in the atom were constantly radiating energy, as the theory suggested, then sooner or later an electron would lose all the energy charge keeping it in orbit and fall into the nucleus. If this happened then the matter which composed the atom would lose its nature and become something else. There could be no material stability.

This problem plagued and intrigued Rutherford and his colleagues. One of these people was a young Danish scientist named Niels Bohr. Bohr suggested that electrons in an atom occupied discrete energy levels and an atom only emitted radiation when an electron changed to a lower energy level; that is, moved closer to the nucleus. It made sense that electrons could only do this if they had somehow received enough energy to have already moved away from the nucleus and out of their discrete energy level. The sequence was simple: an atom received energy to move an electron to a higher energy level further from the nucleus; but this new level was not stable, and the electrons wanted to return to a lower, stable level. When the electrons returned to their stable level, they gave off most of the energy they had received when they moved to the higher level (Figure 2.4).

The Bohr explanation was very effective in describing the way a hydrogen atom worked, but when it was applied to any other atom, it worked less well. When atoms containing many electrons were examined it was clear that physical laws governing particles were not universally applicable. A new, more sophisticated approach was necessary. In 1926 Erwin Schrödinger introduced an elegant formula describing the behavior of the electron in a hydrogen atom. This formula introduced *quantum mechanics* and postulated that an electron acted more like a cloud of charges than a spinning particle. There were characteristics of charge clouds such as shape, location, and spin which could be used to describe the particular energy pattern of an atom. Instead of electron orbits these energy patterns were called *atomic orbitals.* They could not be precisely defined or located at any one particular time but could be predicted according to probabilities. Our basic model of the foundation of matter was now a nucleus with a positive charge and mass, surrounded by clouds of negative charge

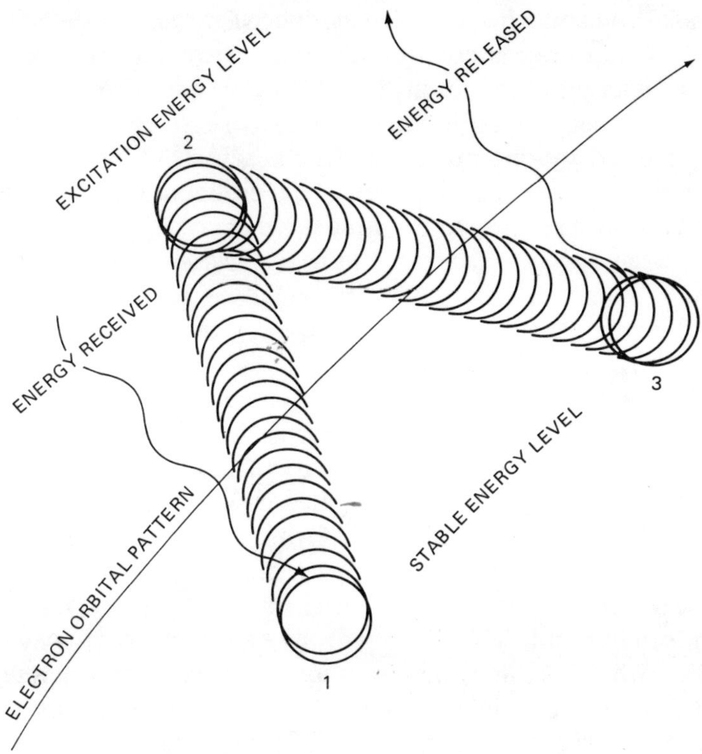

Figure 2.4 Electron energy transfer between energy levels

in predictable patterns of shape, distance, and spin. These atoms existed in energy states with the lowest state being the most stable.

We can no longer say that an atom is composed of particles of matter of the sort we visualize when we think of rocks or lumps of coal. We can only say that atoms are composed of conditions of energy.

There are three subatomic conditions of energy which compose atoms. The nucleus of the atom is composed of a dense mass with a positive charge which spins in a particular direction. This condition is called a *proton.* In addition to protons the nucleus of an atom may contain *neutrons* which have mass but no charge. The nucleus of an atom is thought to be spherical and about 10^{-11} mm in diameter. About the nucleus exist the electrons which are though to be a cloud of negative electrical charge. However, we often refer to them as very small particles at a point in space which can never be exactly located.

It was Thomson who provided evidence for the existence of the elementary condition of energy called the electron. A free electron (cloud of energy) exerts an influence on the space around it. This influence is called a charge and is equal in size to the influence or charge exerted by one proton, even though a proton is many times more massive than an electron. The electron has been assigned a negative charge (-); a proton, a positive charge (+). There is attraction between opposite charges, but like charges repel each other. The mutual influence of charges pervades all of space and varies from place to place depending on the proximity and number of electrons and protons.

In many situations the mutual attraction and repulsion of electrons and protons is in a dynamic balance with other forces resulting from motion. We have seen that the centripetal forces of electrons, which seem to be revolving in regular patterns around protons and neutrons, counteract the attraction of the protons and electrons and stabilize matter. This balance between motion and charge is not constant however. We have noted that electrons move about the nucleus of an atom in atomic orbitals or clouds of charge. We may also think of these orbitals as existing in layered patterns of discrete energy levels called shells similar to the Bohr model. In the outer shell of most atoms there is a greater degree of movement of electrons. These electrons are farther from the nucleus and come under the influence of energy which exists outside their own molecular structure. The activity of electrons in the outer shells of molecules creates both chemical reactions and electricity.

When electrons in a particular molecule receive energy from photons or other quanta of energy, they may be excited enough to move to a higher energy level (shell) relative to their nucleus. They may even be freed completely, in which case they may move to a neighboring atom or molecule and exert their quantum of energy, received from an outside source, to take the place of another electron. When this transfer of energy happens the supplanted electrons receive the same quantum of energy and are freed and move to supplant an electron in another molecule. This process of energy transfer may take place throughout the entire boundary of a particular molecular structure. We call this transfer an *electrical current* (Figure 2.5).

Some materials have outer electron shells in which the electrons are

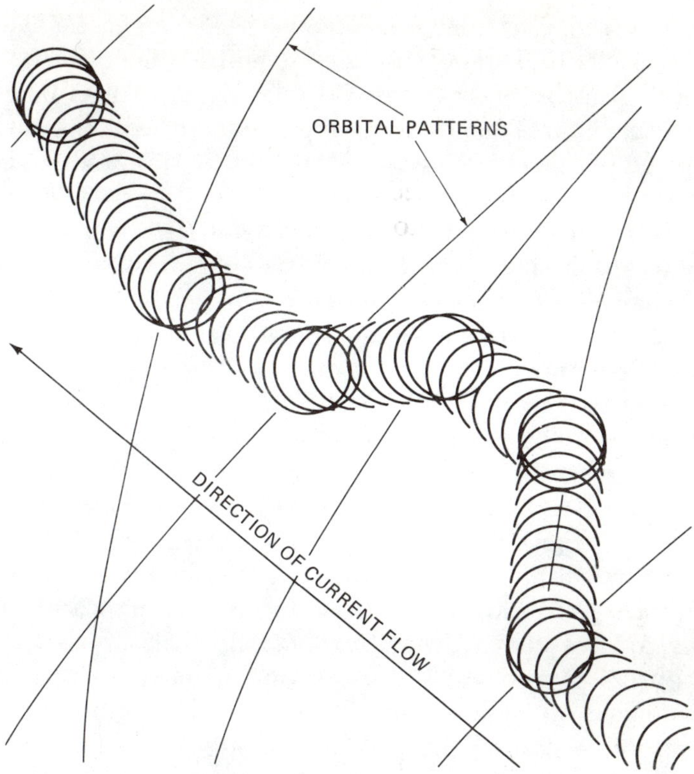

Figure 2.5 Electrical current

very loosely held. Electrons in these materials are easily set free by external energy. These materials are referred to as good conductors of electricity and include copper, silver, and aluminum. When electrons in the outer shell of a molecule are held tightly the material is not a good conductor of electricity and is called an insulator. Materials such as wood and rubber are in this category.

There are many forms of energy which can excite the electron in a molecule. We have seen how electrons can be excited by photons, but the most common method used today to excite electrons and create currents of electricity employs the motion of a conductor through a magnetic field.

An electrical field is a property of space between energy charges. When an electron energy transfer takes place a magnetic field is also produced and an electromagnetic field exists. A magnetic field differs from the electrical field because it exists outside the material through which the electron energy transfer (electrical current) takes place. This external charge surrounding the conductor (the magnetic field) is produced by the current and the relative motion, and it circles the conductor perpendicular to the current flow (Figure 2.6).

The electrical current creates a *magnetic field* when it is moved relative to an observer; however, the two phenomena are part of the same system. If a current can produce a magnetic field, then a magnetic field can produce a current. This is exactly what happens when a conductor is moved through a magnetic field. The production of electrical current by moving a conductor through a magnetic field is the principle of the electric generator—the way most electrical current is now generated. There are many sources of mechanical energy which can be used to move a conductor through a magnetic field and run electrical generators. The one which is used most at present is the steam driven turbine. Steam to drive the turbine, which is essentially an elaborate paddle wheel, is generated by boiling water. The heat for this process comes from burning oil, coal, or gas. However, other sources of mechanical energy such as moving water and wind do not depend on combustion. We will consider some of these noncombustion mechanisms in detail in chapter 3.

We have examined some of the concepts which support our basic understanding of light and electricity. These manifestations of energy are fundamental to a solar hydrogen economy, and they are intimately related to hydrogen itself.

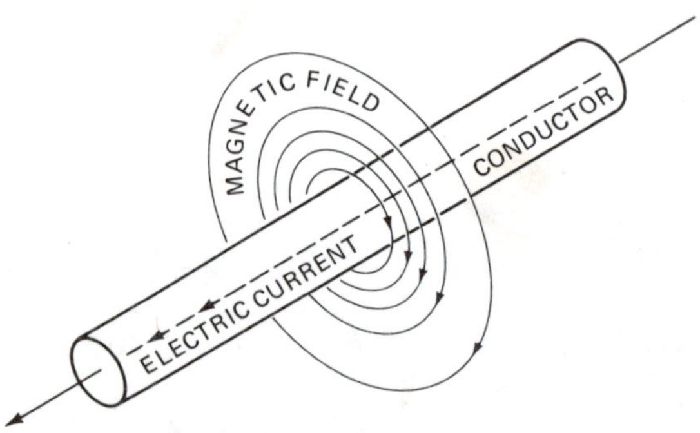

Figure 2.6 Current-induced magnetic field

HYDROGEN

The basic form of stable matter is the molecule which is composed of atoms of the same element or of different elements. When different elements combine to form a stable molecule they are called compounds. The process of combination is called a chemical reaction and results from the electrical charge relationship in atoms.

The least complex, least massive, and most abundant element in the universe is hydrogen. The basic hydrogen atom is composed of one proton and one electron (orbital). When not combined with another element the hydrogen molecule exists as two atoms bound together by shared electrons. There are two states of the hydrogen molecule, *ortho* and *para.* In an ortho state both of the nuclei in the molecule spin in the same direction; in a para state they spin in different directions. The para state is the higher energy state. Most of the free hydrogen is a mix of the two states in a ratio of 4 to 1 with the ortho (lower state) being the most common. In addition to two energy states, hydrogen can exist in three forms which are determined by the number of neutrons in the nucleus. Ordinary hydrogen has no neutrons and is called *protium.* Two additional *isotopic* forms (refers to the number of neutrons in the nucleus) also exist. The first of these has one neutron in addition to the proton and is called *deuterium.* The second has two neutrons and is called *tritium.* The isotopes of hydrogen are much less common than the ordinary form. One out of every 5,000 hydrogen atoms is a deuterium atom and one out of 10,000,000 is tritium. Tritium is radioactive, which means that it automatically changes its nuclear structure over time and becomes a different element. It has half-life of about 12.5 years and decays to a form of helium.

Atoms combine with each other to form compounds by negotiating the relative positions of the electrons (negative charges) in their outer shells. Molecules seek stability, and this stability is largely determined by the number of electrons which can exist together to create optimum stability. When fewer electrons are present, molecules seek to fill these gaps with electrons from other molecules willing to yield electrons. Because hydrogen has only one electron and this electron is in a shell that requires two electrons to be stable, it is attractive to molecules that wish to yield electrons. Conversely, it has only one electron in excess of the next lowest energy level. It therefore tends to lose an electron when in proximity with a molecule with a strong attraction for an electron. When molecules negotiate electrons they usually share them and form new substances. Hydrogen is the only element which uses all of its electrons in this negotiation and it combines with more different atoms than any other element.

When hydrogen yields its one electron in a negotiation, it becomes a sort of junior partner in a compound with an element that has a greater electron attraction. In such a case it resembles an *alkali metal* and assumes a positive charge. Hydrogen can also attract an electron when in the proper proximity to an element with a strong propensity to yield an electron. In this case it acts as a *halogen* and becomes negatively charged. When an element yields an electron in a chemical process it is called a *reducer*, and when it takes an electron it is called an *oxidizer.* Hydrogen can act as both, depending on the other element or elements involved.

Sometimes hydrogen is forced to act as both a reducer and an oxidizer at the same time. This happens when the element is placed in a situation in which it either attracts or yields an electron with a particular element but is also attracted by another element. When hydrogen is attracted simultaneously by two atoms, *hydrogen bonding* occurs. Only one strong bond exists, but there is a powerful additional attraction to another atom. A good example of this relationship is water.

Oxygen is very electronegative; it attracts electrons from hydrogen when it interacts with it to form water (H_20). The attraction of oxygen for the hydrogen electron is so strong that the hydrogen atom is left with a positive charge due to an excess of energy in the nucleus because oxygen has captured the balancing negative charge. This positive charge is so strong that it forces the hydrogen atom to act as an oxidizer when confronted with another atom of oxygen. Each atom of hydrogen in a water molecule is attracted to oxygen in another molecule. This *hydrogen bonding* is what makes water so stable. In a sense oxygen and hydrogen are held together by a double force. Water molecules don't release each other easily due to this bonding; thus, water boils at a much higher temperature than it would if bonding did not exist. If hydrogen did not have its unusual atomic structure, water would not exist in a liquid state at temperatures and pressures which support life as we know it.

In a free state, under normal conditions on earth, hydrogen is an odorless, colorless, tasteless gas. It is the lightest known substance, having a density of 0.08987 grams/liter at 0° and 760 mm pressure. It is about one-fourteenth as dense as air.

Hydrogen is the most abundant element in the universe and is thought to be the source of all other elements through the process of nuclear fusion in stars. Because hydrogen is so chemically active, free hydrogen is relatively rare on earth. It exists primarily in combination with oxygen as water or *hydroxyls,* and with carbon, oxygen, and trace elements as organic compounds. Free hydrogen is relatively stable at room temperatures. It requires heat or some other form of energy to combine with most other elements. Hydrogen and oxygen exist together at room temperature without combining. However, when sufficient energy is supplied they combine to make water; and when they combine they release energy, thus enacting a fundamental process in the solar-hydrogen energy economy. This energy can be released either in the form of heat or electricity.

When hydrogen and oxygen are put together and sufficient heat is added, they combine vigorously and release a great deal more heat. The temperature of combustion can reach 3000°C. There is no light released in this combustion; the only product in addition to heat is water. Combustion will take place over a wide range of hydrogen/oxygen ratios. Depending on the heat of combustion, as little as 4% by volume to as much as 75% by volume of hydrogen will combine with oxygen to produce water. The process will also take place over a wide range of temperatures.

If hydrogen and oxygen are placed together in an electrolyte they will combine to produce water and electricity. This process is basic to the operation of a fuel cell (Figure 2.7).

Just as hydrogen and oxygen can combine to make water and release energy, so water can be taken apart to produce oxygen and hydrogen. The same energy which is released on combination of these elements must be supplied to separate them. For example, water will dissociate into hydrogen and oxygen if sufficient heat is added. It will also dissociate if a current is passed through it. Since water is not a

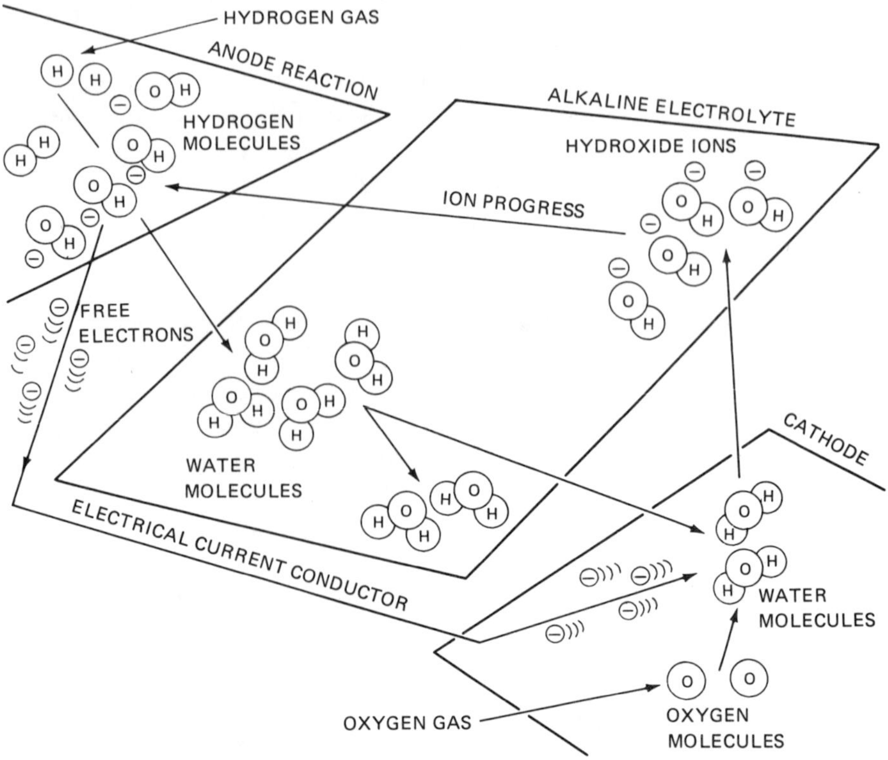

Figure 2.7 Principles of operation of a fuel cell Using an Alkaline electrolyte

very good conductor of electricity, other elements must be added to it in order for the current to flow.

When electrical energy is added to water, the molecule separates and each element carries the electrical charge, an *ion,* responsible for the combination. In the case of hydrogen, the charge is positive; oxygen carries a negative charge. Upon separation of a water molecule, the hydrogen ion is attracted to the negative pole of the electrical field and collects on the *anode.* The oxygen is attracted to the positive *cathode.* These charged molecules (ions) become electrically neutral when they yield their charges and complete the electrical circuit through the liquid. They may then be collected as free, elemental molecules (Figure 2.8).

Through the process of chemical dissociation and reassociation,

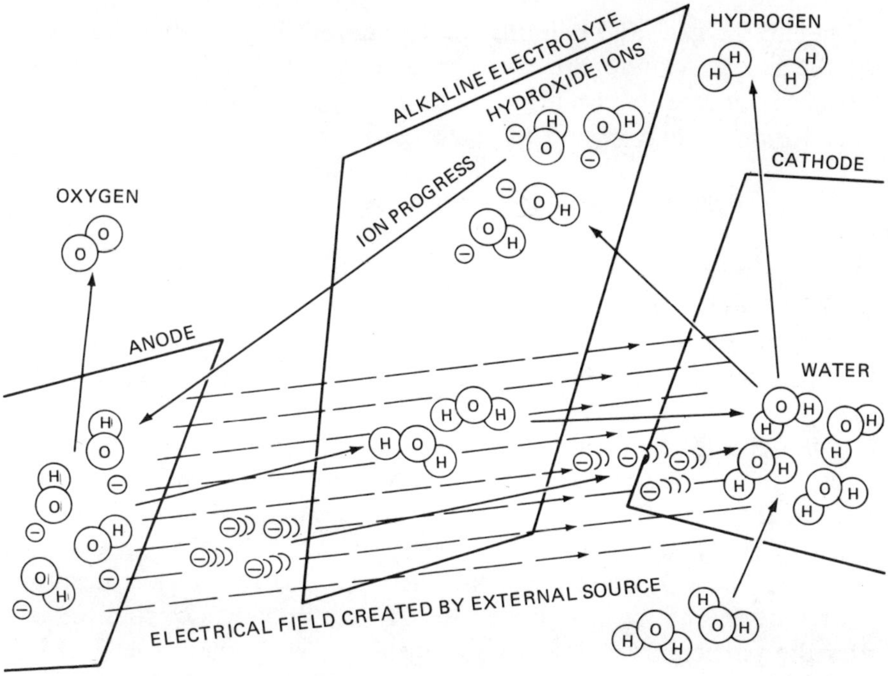

Figure 2.8 Principles of water electrolysis in an alkaline electrolyte

hydrogen provides an efficient and effective energy storage system. Unlike other energy storage systems such as hydrocarbons, the hydrogen system is rapid, reversible and clean. When combined with sunlight as a source of energy this hydrogen storage system results in an energy process which is virtually inexhaustible, globally available, and environmentally beneficial.

THE SOLAR-HYDROGEN CONNECTION

Human beings have long speculated about the nature of the sun's energy. Since the sun is a source of heat it made sense to think of it as a great fire in the sky. As the body of scientific knowledge increased however it became more difficult to explain how a process of combustion could create such an intense source of energy. In 1939, H. Bethe proposed that the energy of the sun was a result of a series of reactions which involved the nuclei of atoms of hydrogen, carbon, nitrogen, oxygen, and helium. He formalized the idea that the source of the energy of the sun was a process by which hydrogen atoms fused to become helium atoms. The energy released in this process is called *binding energy* and the basic formula is:

$$4(H^1) \longrightarrow {}_2He^4 + 2\,({}_1B^\circ) + 24.7\ MEV$$

The release of heat and light through this process is great; however, the conditions necessary to recreate it under controlled conditions on earth are expensive and dangerous. Of course, the process has been duplicated under less than controlled conditions in the thermonuclear bomb (hydrogen bomb).

Solar energy is a result of the way the hydrogen atom negotiates its energy in an environment of high temperatures and pressure in the core of the sun. This energy is the primary source of energy on earth through one process, photosynthesis, the basic chemical process of green plants. For most practical purposes it is the only continuous source of energy we have access to. The combination of solar energy and the particular chemical relationship of carbon, oxygen, hydrogen, nitrogen, and other elements on earth is responsible for that delicate and extremely sensitive life system of which human beings are a part. All of the biomass (the accumulation of living and non-living products of the life system) is a result of solar energy.

Rather recently in the tenure of human beings on this planet, it was discovered that a process of rapid oxidation of the elements in biomass (primarily carbon) such as wood and grass could be created artificially. This rapid oxidation was called fire and it is *exothermic,* which means that heat is produced. The discovery and control of this process gave humans the ability to live comfortably in environments which had been closed to them. Ultimately, the entire planet was acceptable to human life because life-supporting energy could be transported anywhere by the use of fire.

From the moment of the discovery of fire human beings were involved in two great energy enterprises, the first to find something to burn and the other to find more uses for fire. These two great enterprises still dominate human society as we search for more fossil fuels and expand its uses.

Fuel for fires was plentiful for a short period of human history, until most of the great forests were consumed. In that time human society had come to depend so much on fire that other fuels had to be found if society were to continue. In the search for combustibles we turned to the past. Having consumed much of the easily available living biomass, the long-dead, geologically compressed remains of past life became the primary fuels for the human energy system. First, coal was used as a primary combustible; then, as a result of a number of social and technological events, oil became the main combustible. Oil and coal were convenient fuels for fire-making because they were readily available and relatively easily handled once technological routines were established. It is clear however that the most easily extractable oil and coal have already been used. It will require energy to use fossil fuels as time goes on. Unless there are changes, the cost of exerting this energy will be reflected in the general world economy and in the social condition of the human species until the cost is too great a burden for society to bear.

There are other options to the continued dependence on carbon combustion as the primary energy process. One of these is the nuclear process. We have noted that the basic energy of the sun is a result of a nuclear process called fusion, but we are most familiar with the nuclear process called fission which is used in our nuclear power plants.

Nuclear processes release large amounts of binding energy, the energy which holds the nucleus of atoms together. At present we use binding energy in the form of heat to boil water to make steam which turns turbines in order to move wire conductors through magnetic fields to produce a current. This current is then transported over power lines for use in homes and in industry. However, the binding energy of atoms is not released only as heat. Packages of intense energy are also released in the form of nuclear radiation. This radiation yields its energy on impact; when it interacts with living cells the energy is often too great for the cell to continue to live.

There are many analysts who believe that this radiation can be controlled and that nuclear energy is the solution to the problems posed by the limitations of the carbon combustion process. The problems posed by nuclear energy however are staggering and their solutions, if possible at all, appear to be extremely expensive. One serious problem is that nuclear particles continue to be emitted long after all usable heat energy has been extracted from the process. Radiation can in some cases continue for thousands of years. The question of what to do with a used nuclear power plant and with used nuclear fuel has not been answered satisfactorily. It may never be.

Another possible solution to the problems of the carbon-combustion process is the solar-hydrogen process. This process relates the nuclear energy of hydrogen with the chemical energy of hydrogen. The means through which these energies are related is the sun. This process does not necessarily require combustion and it can exist independently of organic processes on earth.

The fusion process in the sun releases energy which is radiated in quanta or discrete energy packages. We have seen that as these quanta of energy impact an atom, they move its electrons to a higher energy level, or release them. When these excited electrons return to a stable state they release most of the energy they have received from the photon. The energy they release can be in the form of heat, chemical energy, or electricity depending on the conditions which have been established. The heat energy which is released on earth by the quanta of solar energy is responsible for the temperature differentials which make wind blow, water evaporate and condense, ocean water circulate, and all aspects of the weather. These manifestations of solar energy involve motion that can be converted into electricity. Wind generators, hydropower generators, and ocean thermal generators are technological devices that convert solar thermal energy to usable forms.

We have seen that the photon is the basis of photosynthesis, but we have also noted another phenomenon. When a photon impacts on an electron of a metal, and the energy level of each is compatible for the process, the electron is raised to a high energy level and can transfer a quantum of energy. Recall that this phenomena is called the photoelectric effect. If the energy transfer created by the electrons in this process is organized to flow in a controlled direction, then a current of electricity can be produced. This process is called photovoltaics, the direct conversion of photons into electricity.

The electricity which results from solar energy, from the use of either motion or photovoltaics, is not a result of heat generated on earth. It is available as long as the sun exists and human beings have the skill to retrieve it.

One of the serious problems with electricity is that it is expensive to transport. We have seen that when an electrical current exists a magnetic field is produced around it. Part of the electrical energy which creates a current is used in the magnetic field. In addition, some of the electrical energy which creates a current makes heat in the conductor. Both of these conditions are energy expensive. Furthermore, the longer the conductor (transport system) the more expensive it becomes. Electrical energy which is transported over long wires loses much of its energy. There is however virtually no inherent loss of energy when hydrogen is transported through pipes. Electrical energy can easily be turned into hydrogen and vice-versa.

The conversion of solar energy into hydrogen can be accomplished by heat, chemical process, or electricity. Hydrogen produced by any of these means stores solar energy and can be used to transport it efficiently over long distances. After storage and transportation hydrogen can be used as heat through combustion or it can be directly converted into electricity.

The remainder of this book is devoted to a description of the elements of an energy economy based on the solar-hydrogen connection. We will also examine the social and economic problems associated with the evolution of such a model.

The solar-hydrogen energy economy is based on the conversion, transportation, and storage of solar energy. The medium through which the transportation and storage takes place is hydrogen. While there are a large number of technologies which could be involved in these processes, only a few are projected to be the foundation of the solar-hydrogen system in the near future.

Fundamental in the solar-hydrogen system that we will consider are four interdependent processes: the conversion of sunlight to energy, the conversion of solar-generated energy into hydrogen, the storage and transportation of energy as hydrogen, and the conversion of hydrogen into energy.

Conversion of sunlight into usable energy can be accomplished in four fundamental ways. The present carbon combustion system is based on the conversion of sunlight into biomass through photosynthesis; the biomass is then burned to produce heat which is used to provide human comfort and survival or mechanical energy. The mechanical energy can provide transportation, or it can run generators to produce electricity. We are all familiar with most of the technologies involved in this system. Through the production of toxic wastes, most of the elements of the carbon-combustion system are probably in opposition to the long-term survival of the human species. There are some aspects such as coal gasification which can be made relatively safe and are important in the transition to a solar-hydrogen system. There are other aspects of the biological conversion of sunlight into energy such as biophotolosis which offer long-range potentials for hydrogen production.

Another way sunlight can be converted into electrical energy is through its effect on the relationships which exist between the earth, water, and atmosphere. This complex process involves the movement of vast amounts of materials over the surface of the earth. The wind, the flow of the rivers, and the waves at sea all are part of this process. Some of this motion can be used to run electrical generators. Hydroelectric plants are common and wind generators have been used recently in greater numbers. This form of solar energy conversion is an important aspect of the solar-hydrogen system.

The heat of the sun can also be used onsite or converted into electrical energy. The intensity of the solar heat which reaches the earth can be increased through refraction or reflection. When light moves through any medium, such as a glass lens, it is refracted or bent; if this bending is controlled, light can be made to focus. Through concentration of light, more heat is released in a smaller area. We can also concentrate light by controlling the angles at which it reflects. Again the intensity of the light, and therefore the heat released on a surface, is greatly increased by this concentration of light. Solar heat which has been created by these means can be converted to mechanical energy through steam and used to run electrical generators.

Finally, sunlight can be converted directly into electrical energy in the photovoltaics process—an important, relatively new, and developing technology. It is one of the major components of a solar-hydrogen economy and will be considered in detail.

The conversion of solar-generated energy into hydrogen can be accomplished through the dissociation of water into its elements in three basic ways: heat dissociation, chemical dissociation, and electrolysis. The first two of these methods appear to have important long-range applicability but require considerable research and development. Electrolysis will be considered at length as an available and important part of a solar-hydrogen system.

Conversion of hydrogen into usuable energy can be accomplished by combustion or in fuel cells. Combustion of hydrogen to power internal combustion and jet engines is a practical and essential first step in the implementation of a solar-hydrogen system. However, the long-range future of the system probably depends on the direct conversion of hydrogen to electrical energy. This process is accomplished through reverse electrolysis. The mechanism for accomplishing this conversion is an electrochemical devise called a fuel cell.

We will examine these three major conversion processes, sunlight to usable energy, energy to hydrogen, and hydrogen to energy in the following chapters.

REFERENCES

Arnold, D. R. et al. *Photochemistry: An Introduction.* New York: Academic Press, 1974.

Bailan, J. C., Jr. et al. *Comprehensive Inorganic Chemistry.* Oxford: Pergamon Press, 1973.

Baker, Adolph. *Modern Physics and Antiphysics.* Reading, Massachusetts: Addison-Wesley Publishing Company, 1970.

Barltrop, J. A. and J. D. Coyle. *Principles of Photochemistry.* Chichester: John Wiley and Sons, 1975.

Capra, Fritjof. *The Tao of Physics.* New York: Bantam Books, 1976.

Campanion, Audrey L. *Chemical Bonding.* New York: McGraw-Hill, Inc. 1979.

Cox, Kenneth E. and K. D. Williamson, Jr., eds. *Hydrogen: Its Technology and Implications.* Vol. 1. "Hydrogen Production Technology." Cleveland: CRC Press, 1977.

Cox, Kenneth E. and K. D. Williamson, Jr. (eds.). *Hydrogen: Its Technology and Implications.* Vol. 5. Boca Raton: CRC Press, Inc., 1979.

Devlin, Robert M. and Allen V. Barker. *Photosynthesis.* New York: Van Nostrand Reinhold Company, 1971.

Dickson, T. R. *Understanding Chemistry: From Atoms to Attitudes.* New York: John Wiley and Sons, 1974.

Eisenberg, D. and W. Kauzman. *The Structures and Properties of Water.* New York and Oxford: Oxford University Press, 1969.

Franks, Felix, ed. *Water: A Comparative Treatise.* Vol. 1. "The Physics and Physical Chemistry of Water." New York-London: Plenum Press, 1972.

Joeston, Melvin D. and L. J. Schaad. *Hydrogen Bonding.* New York: Marcel Dekker, Inc., 1974.

Johnston, Robert M. et al. *Elements of Applied Thermodynamics.* Annapolis: Naval Institute Press, 1978.

Kabbe, Fred and Lois Kabbe. *Chemistry, Energy, and Human Ecology.* Boston: Houghton Mifflin Company, 1976.

Kamen, Martin D. *Primary Processes in Photosynthesis.* New York: Academic Press, 1963.

Kim, S. K. *Physics: The Fabric of Reality.* New York: Macmillan Publishing Co., Inc., 1975.

Krenz, Jerrold H. *Energy Conversion and Utilization.* Boston: Allyn and Bacon, Inc., 1976.

Manahan, Stanley E. *Environmental Chemistry.* Boston: Willard Grant Press, 1975.

McCarty, Robert D. *Hydrogen: Its Technology and Implications.* Vol. 3. "Hydrogen Properties." Cleveland: CRC Press, Inc., 1975.

Meinel, Alden B. and Marjorie P. Meinel. *Applied Solar Energy: An Introduction.* Reading, Massachusetts: Addison-Wesley Publishing Company, 1974.

Mellor, J. W. *A Comprehensive Treatise on Inorganic and Theoretical Chemistry.* London: Longmons, 1963.

Morrison, Robert Thornton and Robert Neilson Boyd. *Organic Chemistry.* Boston: Allyn and Bacon, Inc., 1974.

Murrell, John N. et al. *The Chemical Bond.* Chichester: John Wiley and Sons, 1978.

Okube, Hideo. *Photochemistry of Small Molecules.* New York: John Wiley and Sons, 1978.

Rouse, Robert S. and Robert O. Smith. *Energy: Resource, Slave, Pollutant: A Physical Science Text.* New York: Macmillan Publishing Company, Inc., 1975.

Ryan, Charles W. *Basic Electricity.* New York: John Wiley and Sons, Inc., 1976.

Seliger, Howard H. and William D. McElroy. *Light: Physical and Biological Action.* New York: Academic Press, 1965.

Sherwood, Martin. *The New Chemistry.* New York: Basic Books, 1974.

Theimer, Otto H. *A Gentleman's Guide to Modern Physics.* Belmont, California: Wadsworth Publishing Company, Inc., 1973.

Veziroglu, T. Nejat, ed. *Alternative Energy Sources II, Volume 5, Geothermal Power/Energy Program.* Washington: Hemisphere Publishing Corporation, 1981.

Veziroglu, T. Nejat, ed. *Alternative Energy Sources II, Volume 6, Nuclear Energy.* Washington: Hemisphere Publishing Corporation, 1981.

Veziroglu, T. Nejat, ed. *Alternative Energy Sources II, Volume 9, Conservation, Economics, and Policy.* Washington: Hemisphere Publishing Corporation, 1981.

Weast, Robert C., ed. *CRC Handbook of Chemistry and Physics.* Cleveland: CRC Press, Inc., 1977.

3. SUNLIGHT TO ENERGY

A first step in the process of any energy economy must be energy conversion. While energy is ever present in the universe it is seldom in a form convenient for the support of human civilization. The energy inherent in the molecular structure of wood must be oxidized, or burned, before it can emerge as heat; the motion of water or air must be organized before it can run electrical generators; and the energy of the sun must be directed before it can meet more than the basic needs of human beings. In the short span of human history we have almost filled the earth with people because we have converted, stored, and transported energy in a wide range of combinations.

A number of different techniques can be used to convert sunlight into energy and there are a number of ways of categorizing the technologies involved. Some of these ways are elaborate such as the method proposed by the California Institute of Technology Jet Propulsion Laboratory. Others are simply lists of technologies and applications. In the interest of promoting the greatest amount of understanding in the simplest possible framework, we will consider the conversion of sunlight into usable energy for our model of a solar-hydrogen economy in four broad categories: photoelectrical conversions (photovoltaics), biological conversions, mechanical conversions, and thermal conversions.

All the technologies we will consider are directly related to the production of hydrogen at some stage in the process of implementing a solar-hydrogen economy. Some however will also continue to be used alone as a source of energy without conversion into hydrogen even in a mature solar-hydrogen economy. Photoelectrical conversion, for example, can be used to produce residential energy on home sites with roof-mounted generators. Wind generators are also projected for similar uses. This mix of uses gives the solar-hydrogen economy the unique advantage of being both centralized and decentralized at the same time for different functions.

PHOTOELECTRICAL CONVERSIONS

Photovoltaics is a term first used in 1962 to name the process of the direct conversion of light into electricity. While the term is relatively new, the process has been known for some time, though it has only recently been put to practical purpose.

Photovoltaics is a result of a phenomenon called the *photoelectric effect* which was first clearly described by Albert Einstein in 1905. As we mentioned earlier, Einstein observed that light seemed to exist in discrete packages or photons. A photon of the proper energy level will cause an electron to excite and depart from its stable state. The energy of the photon is a function of the wavelength of light which is directly related to color. Intensity of light is a function of the amplitude of the wave and as we have seen is not a measure of the energy of a photon. Intensity or brightness of light is rather a measure of the number of photons in a unit of time.

When a photon at an appropriate energy level frees an electron, or removes it from a stable state, it gives all its energy to the process and ceases to exist as an entity. The free electron now possesses all the energy of the photon minus the energy used in freeing it. Two conditions have in fact been created: a free electron, and an *electron hole* which is the place where the electron was before it was excited. The free electron will seek stability; when if finds a stable location or state it will release the energy it has inherited from the photon and donate this energy to another electron. This chain of electron energy transfer can create useful electrical current if it is organized (Figure 3.1).

The photoelectric effect can take place anywhere a photon can penetrate, but for convenience we speak of two types of photoelectric effects: external and internal. The internal effect can only happen if photons penetrate the surface of the material. Of course, in order for a photon to penetrate, the material must be at least semitransparent. For the internal process to be efficient the photoelectric effect must take place deeply enough so that the energy imparted to the electron does not move the electron out of the material. The distance an electron moves when freed is called the *diffusion length* which in this case

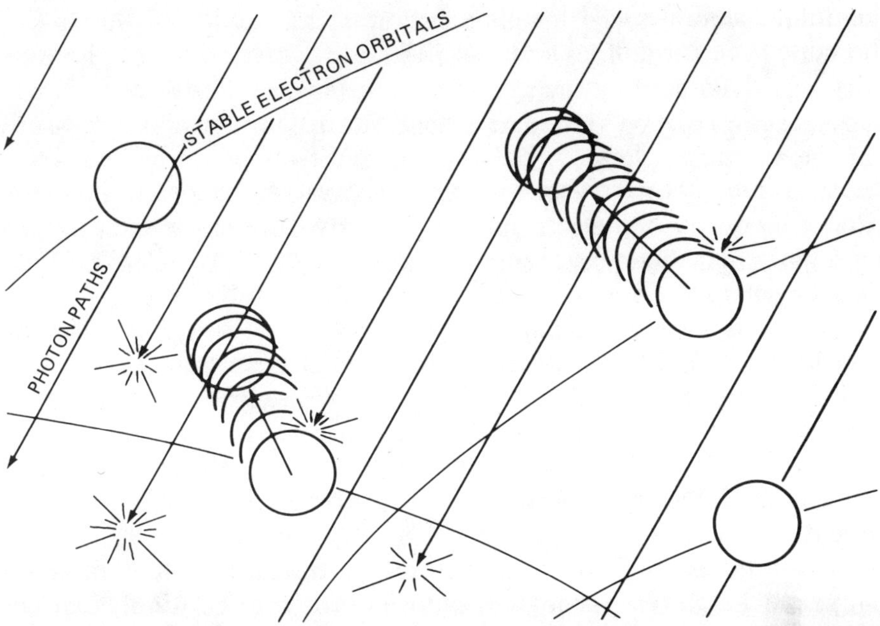

Figure 3.1 The photoelectric effect

should not be longer than the thickness of the material. As we have noted, the electron will return to a stable state. The length of time that it takes to do that is called its *lifetime*. Two determinants of the potential efficiency of the process then are diffusion length and lifetime.

Let us assume that we have a material which will allow the photons, in the wavelengths of visible light, to penetrate deeply enough that the diffusion lengths of freed electrons will not carry them out of the material. Our problem becomes one of capturing the energy which the photon has given the electron. We must somehow intervene and organize the process before the electron can return to a stable state and release its captured energy randomly. One way to create such organization is to encourage the electron to move in a particular predetermined direction.

Conditions can be established in particular spaces which encourage the spaces to take on characteristics different than their environments. These conditions are called *fields:* physical qualities with definite values which remain constant at any point in the space. These physical qualities are manifestations, or results, of the particularly dynamic way the electrons in the space relate to each other. Temperature can be viewed as a field, for instance, and is a measure of the amount of heat. A field can have direction as well as magnitude; in other words the energy relationships can be something other than random. When electrons in a field have an orderly relationship in a particular direction there exists a *vector field*. An *electrical field* is a vector field possessing both magnitude (amount of energy) and direction (the energy exchanges tend to move one way). In order for an electrical field to exist there must be an electrical potential which can be established in several ways, for example, from the buildup of charged particles on large surfaces as happens in a rainstorm and causes lightning. Whatever way this potential is established, it makes all free electrons in its boundaries want to exchange their energy in one direction.

We could establish an electrical field in the photovoltaic material by sending a current through it, but we would then be supplying electricity when we want to generate it. We must rather establish an internal electrical field inherent in the material itself, and this field must be strong enough to encourage the free electrons to move in a particular direction so that their energy can be captured. We can establish such a condition by placing together two substances which possess differing electron spacing characteristics so that one substance desires to lose electrons and the other desires to gain them. A small electrical field can thus be established between these two substances. This field, while weak, can be made strong enough so that an electron, freed by a photon of visible light, is inclined to move from one material to the other before it recombines.

However, we do not want the energy of a freed electron to dissipate itself across the boundary of the two materials. What we must do is establish some sort of *barrier* between the two substances which will not interfere with the electrical field but which will still prevent free electrons from dissipating at the boundary. Once an effective barrier is established we can then provide another, more organized path for electron energy exchange. To create such a path we may attach the two substances together with a conductor. The freed electrons are still subject to the electrical field established by the two materials. As a result they are compelled to move from one to the other, but they are prevented from moving across the boundary by a barrier. Now however they can easily move to exchange their energy through a wire conductor or a liquid electrolyte between the substance.

This first widely accepted commercial photovoltaic technology involved the use of one of the most abundant materials on the earth, silicon. Silicon commonly exists in combination with oxygen as *silicon dioxide* (SiO_2), a form of which is common sand. Silicon is considered a nonmetal, and, after oxygen, it is the most abundant element in the earth's crust. It is called a semiconductor because it does not easily allow the exchange of electron energy, but neither does it completely prevent an exchange. Silicon can exist in crystal or noncrystal (*amorphous*) forms. We are most familiar with commercial silicon in the amorphous form as glass. In the amorphous form, molecules are randomly arranged and have considerable freedom of movement. In fact, glass can be considered a liquid which flows very slowly. Windows in ancient cathedrals are thicker at the bottoms than at the tops because the glass has flowed down as time has passed. While amorphous silicon is increasing in importance as a photovoltaic material, the most common material used at present is the crystal form. We will consider this form first.

One form of silicon dioxide is quartz, which can be found in small granules, in quartz sand, or as large rocks. The molecules in quartz are linked together in regular and permanent patterns. These patterns are called *lattices,* and the overall order of these lattices is called the *crystal structure.* Generally there are many units of crystal structures in naturally occurring quartz.

Converting quartz into high-grade semiconductor material, which is used in the electronics industry, involves essentially three processes: purification, crystal structuring, and cutting.

The purification of quartz involves the removal of oxygen from the compound and the further removal of impurities from the silicon. The first step in the process is the melting of the quartz and the addition of carbon to the melt. The carbon combines with the oxygen and is released as carbon dioxide (CO_2). The remaining material is almost pure silicon—about 99% pure. Any impurity decreases the semiconductor efficiency of the material. It should be noted that silicon is the basic material used in the semiconductor industry and for these uses it must be extremely pure. Fortunately such purity is not required for photovoltaics. Acceptable purity of the photovoltaic material can be obtained at less expense than that required for semiconductors. Separating the requirements of the semiconductor process from the photovoltaic process at the production level is an important step in lowering the cost of photovoltaics.

Once silicon of the proper purity is obtained the number of individual crystals must be reduced because electron energy is dissipated across the boundaries of crystal structures. The fewer the crystal structures the greater will be the efficiency of the material. If the pure molten silicon is cooled slowly it can be made to grow into a single crystal. The most common method for growing single-crystal silicon is called the *Czochralski Process.* In this process a seed crystal of pure silicon is introduced to the molten silicon. The molten silicon is cooled slowly and grows according to the pattern provided by the seed crystal. As the new crystal grows it is slowly drawn out of the crucible. What emerges is a cylindrical ingot of single-crystal silicon up to six inches in diameter and a yard long. The most common size however is about two inches by eight inches.

The process of forming this single-crystal silicon takes considerable time and requires great precision. The ingot is usually cut into very thin wafers, at high cost, and is used in the electronic industry to make microchips for computers, in electrical circuits as transistors, and for a range of other related uses. The silicon produced in this manner also has properties which make it a good photoelectric device. One of these properties is a molecular structure which allows for the creation of a permanent internal electrical field.

A molecule of silicon has four electrons in its outer shell. In order to be complete this outer shell needs eight electrons; it can obtain the equivalent of the required additional four electrons by sharing electrons with other molecules. It will share if the electrons have enough additional energy to move into shared patterns. The needed additional energy can be added as heat, and this happens when silicon is melted. In the molten form the electrons have the mobility to bond together or share energy. Each electron in a molecule of silicon links with an electron in another molecule. This bonding means that all molecules are linked together so that each molecule's four electrons are paired with four electrons from four other molecules. The shape of this arrangement is a *tetrahedron* which results in a cubic crystal structure. We have seen that if the molten silicon is cooled slowly, and if the basic crystal pattern has been introduced in a controlled fashion, then this crystal structure will continue to grow throughout the entire material and create a single crystal. If on the other hand the material is cooled rapidly the crystal growth does not take place in a regular fashion and an amorphous form develops. Various stages of crystal formation can be achieved ranging from the random amorphous form to a single crystal. These stages are called *polycrystal* forms. When the silicon is cool the molecules are locked together in whatever form they arranged for themselves during the cooling process.

It is possible to introduce enough energy to excite an electron in a molecule of silicon enough for it to break out of its pattern. The amount of energy which is donated to an electron to free it, and which it possesses after it is freed, is represented by its *conduction band* which is the range of energy needed for an electron to create an energy exchange which can result in an electrical current. For silicon this energy is about 1.1 electron volts.

When an electron receives enough energy to reach its conduction band and is free to move and exchange energy necessary to cause electrical current, it leaves its place empty in the crystal arrangement. This empty place, an electron hole, has a net positive charge and desires another electron. A neighboring electron can move to fill this hole if it has received energy equivalent to its conduction band. We have seen that this movement of electrons into electron holes is the energy transfer which is electrical current.

An electron of silicon can receive energy necessary to move it into its conduction band in a number of ways. One is through the action of light, the photoelectric effect which was discussed earlier. When a photon of light hits crystal silicon it usually penetrates the surface deeply enough to yield its energy to an electron of silicon so that the freed electron has a diffusion length which allows it to transfer its energy to another electron in the material. We mentioned earlier that the energy an electron can receive from a photon is a function of both the electron and the photon. The necessary energy which an electron of silicon must receive to be moved into its conduction band, and be set free to move, is equal to the energy in the wavelength of 1.1 microns, which is in the infrared part of the spectrum. Wavelengths longer than 1.1 microns are not used in the energy exchange process. We can therefore expect much of the photon energy in the visible spectrum to be used in the process of freeing an electron of silicon. However the energy of light in excess of 1.1 microns is turned to heat in the process, and some is also reflected off the surface. Due to these losses, which result from heat transfer and reflection, only about 20% of the light which strikes the silicon can be used to free electrons, and this is the maximum efficiency of a photovoltaic cell made of silicon.

The job of converting light into electricity in silicon is only half done once electrons are freed. If no other conditions are created the freed electrons will recombine with other molecules or leave the surface of the material in an unorganized fashion and all their energy will be dissipated. As we mentioned earlier we must organize this electron energy so that we can use it as electric current. This organization is achieved by creating a permanent electrical field in the material by adding other elements to the silicon structure which create an imbalanced charge.

A small number of molecules of another substance can be introduced into the silicon and locked permanently in the crystal structure. If this substance has more than the four electrons which silicon has in its outer shell, the extra electrons will have no place to bond. Because they cannot combine with other electrons to complete a stable shell, they will exist in a free state, and the crystal formed by the silicon and the new substance will have a net negative charge due to the excess of free electrons.

Conversely, if a substance with fewer than four electrons in its outer shell is permanently introduced into the silicon crystal, then a net positive charge is created due to the electron holes which result from a shortage of electrons in combination.

The two different crystal substances are placed together—the crystal with the excess electrons (the net negative charge) is placed against the crystal with the electron holes (the net positive charge). As a result of this placement an electrical field is created across their boundaries. No current flows however because the electrons are permanently locked together in the crystals.

The process of introducing molecules of the substances into the silicon is called *doping;* common materials used are phosphorous and boron. The particular substance is introduced to the molten silicon and locked into the crystal during the growing process. Phosphorous has five electrons in its outer shell and when locked in crystal form with silicon each phosphorous molecules creates one extra electron. Boron has three electrons in its outer shell; therefore, each boron molecule creates one electron hole or positive charge (Figure 3.2).

If a photon of the proper wavelength (1.1 micron) hits a thin photovoltaic cell, which is constructed of silicon doped with boron on one side and silicon doped with phosphorous on the other, then the electron freed by this photon will attempt to discharge its energy in the direction of the net positive charge, that is, toward the side with the electron holes. These freed electrons will try to cross the boundary of the materials but will be prevented from doing so because the electrical field at the boundary of the two pieces of silicon is a *static electric field,* a sort of electrical no-man's-land called the *cell barrier.* However, we can provide another means for the freed electrons to exert their energy from one side to the other. We can let them exert their energy through a conductor, a silver or copper wire connecting

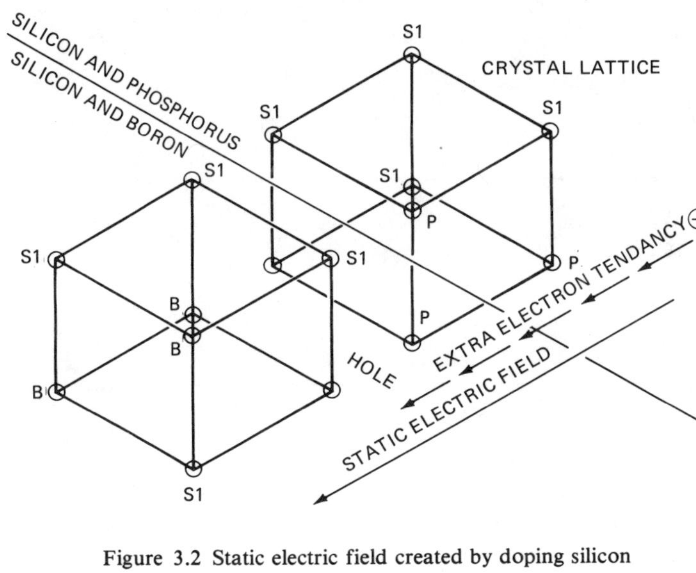

Figure 3.2 Static electric field created by doping silicon

the two sides. We can use the electric current, created by the electrons in this wire conductor, to do any work we wish (Figure 3.3).

While all photovoltaic devices operate according to the principles of the photoelectric effect, there are many ways of creating electrical fields and cell barriers. In addition to the method outlined above, called a *homojunction* because it uses one basic material, other sorts of materials can be faced together. When two dissimilar materials are used such as copper sulfide and cadmium sulfide, a *heterojunction* is created. When a semiconductor, such as silicon is coated with a metal we have what is called a *Schottky barrier*. Another method involves the separation of two layers of a semiconductor with an insulator, called a semiconductor-insulator-semiconductor (SIS) junction. Similarly, the system of a metal and a semiconductor is called an MIS junction.

The theoretical efficiencies of solar conversion for photovoltaic devices constructed of different semiconductors range from 13% for germanium to 27% for aluminum-antimonids; however, for a variety of reasons, it is unlikely that more than 80% of theoretical efficiency will ever be reached (21.6% for aluminum-antimonids). When considering this figure it should be kept in mind that in the generation of electricity from fossil fuel more than 60% of energy can be lost in the

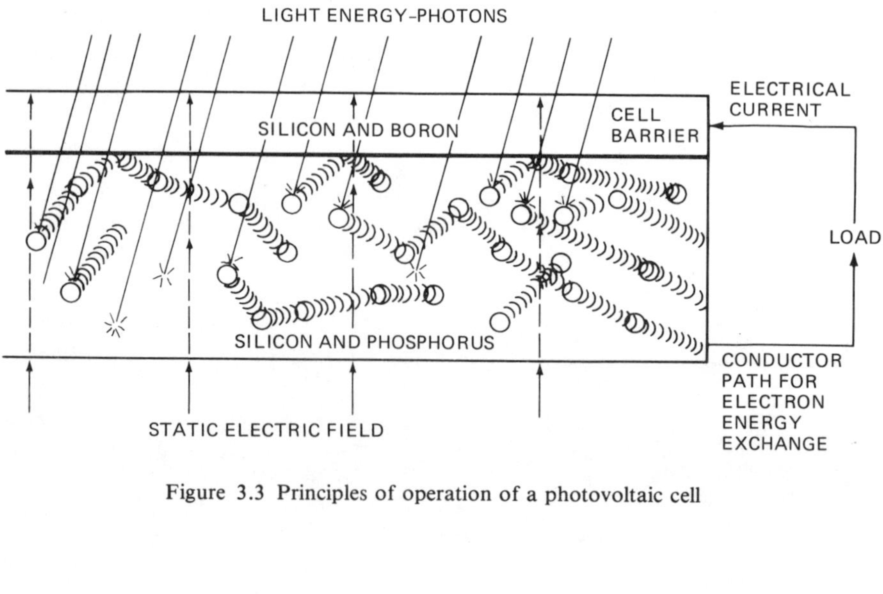

Figure 3.3 Principles of operation of a photovoltaic cell

conversion process and about half of the net energy produced can be lost in transmission.*

While the practical efficiency of solar conversion for photovoltaic devices may never rise much beyond 20%, the amount of electrical current produced by these devices can be improved by increasing the number of photons which impact the surface of the photovoltaic material. An increase in photons can be created by increasing the intensity of the light through magnification of the light either by reflector or by a lens. Such magnification has a dual effect; it increases the number of incident photons per unit of time and also the amount of heat released. The heat however tends to reduce the efficiency of the photovoltaic process so cooling devices must be employed.

Combinations of semiconductor materials with metals and liquids, various types of junctions, and light-concentrating systems involving Fresnel lens and reflectors offer a range of diverse and promising technologies for photovoltaic devices. The most important consideration involved in the implementation of photovoltaic technology does not directly involve increasing the efficiency of photovoltaic devices. It involves instead the efficient manufacturing of such devices. Production of photovoltaic devices has thusfar been so expensive that the technology has not been used except in isolated, unique circumstances. As previously mentioned, single-crystal silicon is the most common material used to make photovoltaic devices. Single-crystal silicon however has been expensive to produce. Since it has been produced primarily in the semiconductor industry in comparatively small amounts, this expense has been easily integrated into the price of the final product. When it is used to make devices for generating electricity however it is used in larger quantities.

Several new production processes are revolutionizing the photovoltaic industry. These new processes span a wide range of construction techniques and promise an equally wide range of applications.

One new system of production which has great promise for residential application has been developed by Texas Instruments In-

* For a good coverage of the principles of various photovoltaic devices as well as relative efficiencies see Paul D. Maycock and Edward N. Stirewalt, *Photovoltaics,* (Andover, MA; Brick House Publishing Co., 1981).

corporated of Dallas, Texas. In this system high-purity, molten silicon is forced through a small capillary opening and the droplets are allowed to free-fall about six feet, forming tiny spheres which range in structure from five grained polycrystals to single crystals. The spheres are remelted in another drop which preserves their shape but improves the general crystal structure. Half of these tiny spheres are then doped with boron and the other half are doped with phosphorous so that half have the *positive* or electron-hole side exposed and half have the *negative* or electron-rich side exposed. The spheres are embedded in a thin sheet of glass and a conduction plane is established on one side of the sheet so that an electrical conductor is established between the positive and negative spheres. The whole sheet is then immersed in an electrolyte, in this case hydrobromic acid (HBr). When the sheet is exposed to light, a current flows between the spheres through the conduction plane and an electrical potential is established between the spheres. The negative spheres thus become cathodes and the positive spheres anodes. The hydrobromic acid is dissociated in the resultant electrolytic process, and hydrogen is released and collected. The hydrogen is stored in a metal hydride and later reconverted to electricity in a fuel cell. The production of the spheres of silicon is reported to result in greatly reduced costs while the efficiency of photovoltaic conversion remains high at about 13%[2] (Figure 3.4).

The Texas Instrument system is applicable to the residential

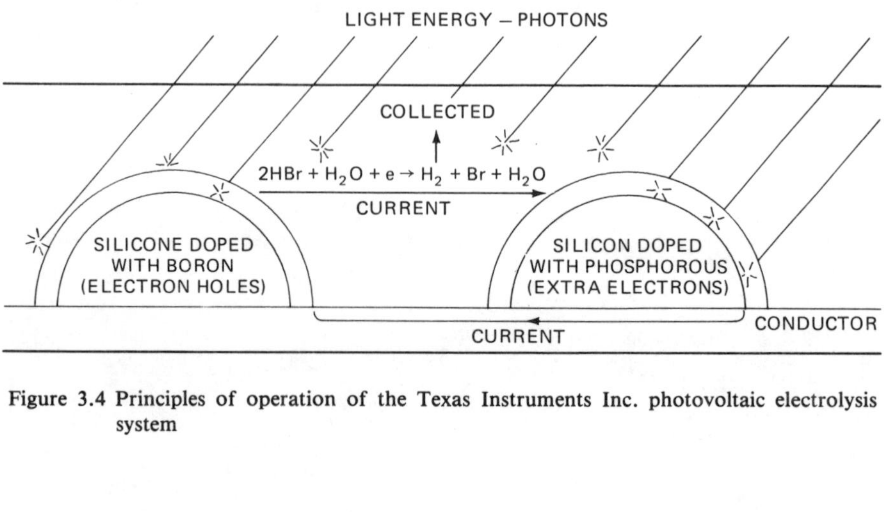

Figure 3.4 Principles of operation of the Texas Instruments Inc. photovoltaic electrolysis system

market because it generally requires no additional support structure to be roof-installed and there is a good match between energy requirements and the area of roof needed for installation. In addition, the system of photovoltaic-material production uses available low-cost materials (SiO), permits lower cost manufacturing techniques, and is called a "forgiving" manufacturing process—because a large number of spheres are produced, a small number of flawed spheres does not significantly decrease the efficiency of the system as a whole. Finally, there is an efficient enough energy payback to make the system very attractive for new home construction in the near future.[3]

The integration of solar electrical generation, hydrogen production and storage, and fuel cell generation of electricity has exciting potential in other new systems. J. O'M Bockris, head of the Texas A & M Hydrogen Research Center recently announced a hydrogen production rate from a photovoltaic system of 13% over a four-hour stable operation period.[4] He has indicated that such a system will produce hydrogen at the equivalent price of $1.00 per gallon of gasoline.

Other new methods of producing photovoltaic devices promise to significantly reduce the cost of solar electricity below the present costs. This reduction cost would provide for a number of applications which are not possible now. One method of production which has been the subject of much recent research uses amorphous silicon in a thin film for the generation of electrical current. There are several advantages to using amorphous silicon in photovoltaic devices. These devices are relatively inexpensive because they use about one six-hundreth of the silicon necessary for a traditional wafer device. They can also be fabricated at a much lower temperature by a glow discharge of silane (SiH_4) on a sublayer of glass. Finally, it is possible to make very large panel areas in this process.[5]

A problem with the use of amorphous silicon has always been its relatively low efficiency of conversion. It has been a widely accepted conclusion that a conversion efficiency of at least 10% is necessary before these devices can be competitive in the production of electricity. In 1982 this 10% conversion efficiency was achieved by scientists at RCA Laboratories in Princeton. Based on the phenomenon that the addition of carbon to silicon increases the optical bond gap of amorphous silicon cells, the RCA scientists achieved an efficiency of

10.1% in a device with a wide-band-gap P-layer of amorphous silicon alloy. The $SI_{1-x}C_x$ P-layers were constructed using a mixture of silane (SiH_4) and methane (CH_4) to which B_2H_6 was added as a dopant. This wide-band-gap P-layer was combined with an i-layer with a relatively long diffusion length and the thicknesses of both layers were optimized.[6]

The head of Photovoltaic Processes and Applications Research at RCA Laboratories has reported an estimation that the production cost of two by four foot panels which develop electricity at 10% efficiency and which are combined to produce 50 MW of power annually can be less than $.50 per peak watt using 1982 dollars.[7] This estimation indicates that the system could have almost immediate application for the production of electricity by public utilities.

While experts disagree about how soon and to what extent photovoltaic-produced electricity will enter the market place in a major way, there seems to be almost universal agreement that this technology will ultimately have great impact.*

The model which we are considering here proposes in increase in the use of photovoltaic electricity to 0.4% of the total energy production in the first stage, a period of 15 years. This increase is great and it may at first glance seem unreasonable; however, it must be realized that recent developments seem to indicate that such an increase is possible. The estimated cost per peak watt of energy produced by photovoltaics in early 1982 was about $9.50. The possibility of producing energy at $.50 per peak watt which has been suggested by RCA is one-nineteenth that amount. This reduction should substantially affect the utility market just as the solar-hydrogen processes should substantially affect the residential market.

The model further proposes that the energy produced by photovoltaics will increase to 25% of total production in a fully mature solar-hydrogen economy.

There are several other new production processes for photovoltaic materials. Most of these processes result in materials with less efficiency of conversion than single-crystal silicon, but they promise to be far less expensive to produce. Among the more promising of these

---

*For an interesting review of the perceptions the photovoltaic potential in 1982 see Don Best, "Point", *Solar Age Magazine,* July 1982.

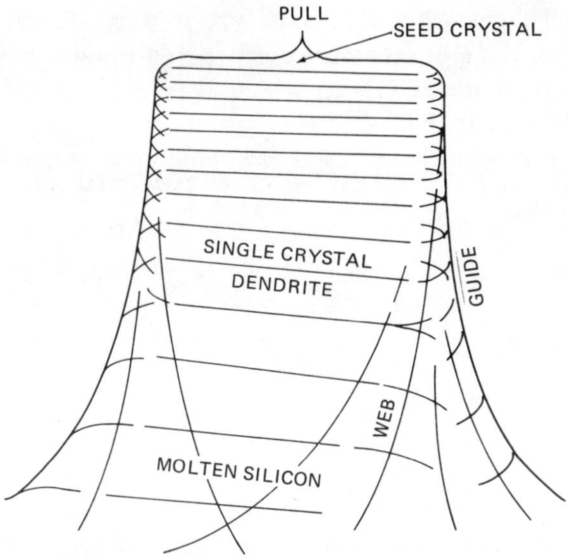

Figure 3.5 Dendritis growth method for producing photovoltaic silicon

processes is the *dendritic growth process*. In this process molten silicon is drawn out of a crucible and grown using shaped dies so that it emerges as a single-crystal sheet which is comparable in performance to that produced by the Czochralski process. Another process, called *the lateral pulling* method, allows the molten silicon to grow into a single crystal as it is drawn from a trough-shaped crucible* (Figure 3.5).

Recently new methods of purifying silicon have been under development which could reduce the cost of solar-grade silicon to one-tenth of their present cost.[8] This reduction in the cost of purifying silicon, when combined with new production techniques, could result in vastly reduced prices for photovoltaic devices between 1985 and 1990.

Photovoltaic devices combined with electrolysis-hydrogen-production systems and fuel cells for the conversion of hydrogen into electricity are an essential aspect of the model solar-hydrogen economy. In our model plan electrical energy produced by photovoltaic devices

*For an important coverage of photovoltaic devices technologies see David L. Pulfrey, *Photovoltaic Power Generation (New York:* Van Nostrand Reinhold Company 1978).

is projected to rise from 0.4% of the total energy production in the T to T + 15 time frame to 25% in the mature solar-hydrogen economy.

BIOLOGICAL CONVERSION OF SUNLIGHT TO ENERGY

One of the most familiar conversions of sunlight into energy is through the process of photosynthesis. In this process carbon dioxide (CO_2) is reduced to carbohydrate (CH_2O) in green plants. Sunlight is the energy source for this process which is fundamental to all life on earth.

Human society's dependence on the biological conversion of sunlight to energy for food and fiber makes the transition to a solar-hydrogen energy economy very important because the continued combustion of biologically produced fuels such as coal, oil and natural gas robs future generations of resources necessary for clothes, drugs, building materials and a range of other materials. Ironically, a final product of the combustion of biological fuels is carbon dioxide, which as we've seen is used by green plants in the process of photosynthesis. Moreover carbon dioxide is produced during respiration and is therefore fundamental to the life process. However, when it is produced in large amounts in a relatively short period of time, it cannot be used rapidly enough in the life process and rises to the upper atmosphere where it remains as a trap for solar heat. Many scientists fear a carbon dioxide catastrophe through what is called the *greenhouse effect* which could drastically impact weather and climate patterns. In addition to carbon dioxide, biological fuels produce a range of potentially toxic materials when they are burned.

MECHANICAL CONVERSIONS

Heat released by photons of sunlight is the energy which powers air and water motion on the earth's surface. The movement of air (wind) is caused by the difference in pressure resulting from the heating and cooling of large air masses. Our changing weather is a manifestation of this movement of air and the evaporation and condensation of water. When water is heated it becomes vapor, which rises and is moved by the wind. As it rises it is cooled and condenses to become cloud cover and finally rain or snow. It moves as a liquid from high points on earth to lower points and finally to the oceans or the deep underground reservoirs. Water in movement across the surface of the earth comprises rivers, streams, and aquifers.

The motion of air and water are fundamental supports of the earth's life system but this motion can be converted into mechanical energy for direct use by human beings. Windmills and watermills have been used for centuries to grind grain and to run the machines of early human industry. In recent years wind and particularly water power have been used to produce electricity in electrical generators.

Water power can be centralized through the use of dams, which cause small streams and drainage areas to converge in reservoirs. Controlled amounts of water are then allowed to escape at great force; this powers turbines and electrical generators. The centralizing of water power through the use of dams has interrupted the natural flow of many major rivers with both positive and negative results. Large dams divert great amounts of water and also change the rate of water flow and depth. Both of these changes have an impact on aquatic life systems. These and other effects of water power systems based on dams must be taken into account when evaluating energy production potential of hydropower.

Ocean waves and tides are also a source of mechanical energy which can be used to generate electricity. In addition to mechanical energy, oceans provide a source of thermal energy which is becoming increasingly important. The difference in the temperature of the ocean at various depths can be used to generate electricity. This form of energy is called *ocean thermal energy conversion* (OTEC).

We will consider each of these major forms of wind and water conversion in this section. Due regard will be given to the fact that these technologies must be employed with consideration to the effects of their use on the life systems which they support.

Wind Conversions

There are three major patterns of wind current. One of these is the pattern of coastal wind. During the day the temperature of the land near a large body of water increases faster then the temperature of the water. Air above the land heats and rises. It is replaced by the cooler air over the water which is heavier. Thus during the day the coastal wind blows from water to land.

During the night the land mass cools more rapidly than the water. Air over the land increases in weight and settles toward the earth. The warmer and lighter air over the water is then replaced by the heavier air from the land. This means that at night the coastal winds blow toward the water.

An even larger pattern exists because there is permanent temperature difference between the equator and the poles. In the warmer areas of the earth, particularly approaching the equator, air rises and is replaced by cooler surface air moving from the poles. This movement is further influenced by the rotation of the earth which forces air movement in the direction of the rotation. The cold surface air moving from the north pole toward the equator tends to twist west while the upper air tends eastward. This condition causes low pressure areas in the northern hemisphere to have air patterns in a counterclockwise direction while low pressure areas in the southern hemisphere have a clockwise wind flow.

Another major wind system is caused by the difference in temperature at different altitudes. Heavier colder air at higher altitudes flows to replace warm air at lower altitudes which moves upward. This system of air movement is varied by the heat of the sun during the day.

As a result of these major patterns of wind, certain regions of the earth are particularly favored with regular and predictable wind currents which can be used to generate energy. Coastal areas, mountain slopes, and plains beneath mountain ranges are among these areas.

The energy in wind is the energy of motion or *kinetic energy*. It must be transformed into motion which suits the desired work. Wind is converted into useful motion for transportation by using sails, probably the oldest form of windpower. It can also be used to turn a shaft; this rotary motion is used for a number of purposes including running electrical generators. Some early wind machines used canvas sails to convert wind energy to rotary power. These sails were attached to the arms of windmills and could be removed or added depending on the force of the wind.

In one form or another, windpower has been an important aspect of the energy economy for almost the entire span of human history, but it decreased in importance when coal and oil came into general use. Windpower has many characteristics which make it an important aspect of the solar-hydrogen economy. Similar to most other forms of solar energy it is renewable, and, though wind machines may be considered a form of visual pollution to some people, they do not threaten environmental quality. The aspect of windpower that we will be most concerned with is the conversion of wind energy to electricity in generators. This energy is then used on or near the site of production, or converted to hydrogen for storage or transportation. While this aspect of windpower is an essential part of the model and will be considered in some detail it should be understood that there are other aspects of windpower which may be of importance in the future, particularly in transportation. For example, efficient new sail designs may make travel and commercial shipping by sailing craft very attractive when fossil fuels are extremely expensive. New sailing craft designs like the Club Mediterranean's large, titanium-hulled, four-masted ship *Mediterranean* equipped with photovoltaic devices may become common in the future.

While windpower is a phenomenon which surrounds us all the time, the principles on which wind machines function are not commonly understood.

The energy of wind is motion, and this motion is a result of differing air pressures between adjacent points in space. This difference in pressure is really a measure of a difference in the number of gas molecules between two points in space. Such differences cannot exist as a static condition. Gas molecules in one point in space will move to fill vacant areas in other adjacent points. The motion of gas molecules is wind and we can measure the speed of the molecules as *wind velocity.*

We can conceive of wind energy in two ways. First, wind is a force of the moving molecules, or air, against a windmill blade or sail. We can also think of it as the motion of a windmill blade or sail from a place of high pressure to a place of lower pressure.

When wind energy is though of as the force of the moving molecules of air it is the *kinetic* energy of air, and we can describe this energy in the following terms:

Kinetic Energy = $1/2pV^2$

p = the density of air (the number of molecules per unit of space)

V = the velocity of air molecules relative to a stationary point on the earth.

Power is a term which describes the amount of energy in a unit of time. In the case of wind, power is the amount of kinetic energy which exists in a specified amount of time.

Most wind energy machines or windmills are horizontal-axis machines. If we ignore certain effects, which are assumed to be of little importance in the actual operation of windmills, we can mathematically describe the power obtainable from a particular horizontal-axis machine. The effects which this description ignores include the kinetic energy imparted to the wind by the rotation of the blades, and variations in wind speed and density during flow. The formula for power in a horizontal-axis windmill then is:

Power = $1/2\, p\pi R^2\, V^3$

πR^2 = the area through which the windmill blades travel

$1/2\, pV^2$ = kinetic energy of air

In order for a windmill to be 100% efficient in using the kinetic energy of the air it would have to use all the motion of the wind and the wind would cease to blow in the area through which the blade moves. This condition obviously does not occur because the wind continues to move toward the downwind side of the blades. The maximum power which can be extracted from the wind exists when the speed of the wind which emerges from the moving blades is one-third of the speed of the wind before it enters the blades. When this fact is combined with the formula for the power of a horizontal-axis windmill, a formula for the maximum power which can be extracted from a horizontal axis machine can be developed. Under optimum conditions the maximum power is:

$$\text{Maximum Power} = \left(\frac{16}{27}\right)\left(\frac{1}{2}\right)\pi R^2\, pV^3$$

This formula tells us that under the best of conditions only 59.3% of the kinetic energy of the wind can be converted into usable power in a given area. Modern windmills convert between 50 and 75 percent of this available energy.

Wind generators can be grouped together for centralized power generation or they can be used individually to provide energy for individual home sites. It is estimated that one gigiwatt (100,000,000,000 watts) of power (roughly the output of a large conventional power plant) could be generated by about 5000 wind machines operating at 75% of maximum power. Each of these machines would have turbine blades of 60 meters in diameter and would operate in an area where the wind blows at 23 mph or better for 25% of the time. Machines of this capacity have been in operation on a test basis for some time. In the early 1940s a wind machine built by Smith-Putnam was rated at 1250 kW in 34 mph winds. This machine had a blade length of 53.4 m. It operated successfully for 1000 hours between 1941 and 1945.[9]

Several large wind generators have recently been placed in operation or are scheduled to be operational in the near future. Three 200 kW NASA/ERDA turbines are being installed as a result of the test of the model erected in 1975 at Sandusky, Ohio. Another, with a motor diameter of 54 m in Ulfborg, West Jutland, Denmark, is rated at 2000 kW.[10]

While centralized power generators using wind are a relatively new possibility, decentralized systems for both domestic and commercial purposes have been used for many years.

Vertical-axis wind machines were used in Persia over 1000 years ago for pumping water and grinding grain. Windmills were used extensively in Europe beginning in the Middle Ages. They were employed because they were less expensive than either human or animal energy—the same reason for the popularity of water power.

The use of windpower in the Middle Ages in Europe predicted an essential problem of modern solar energy; its great availability was incompatible with the economics of monopoly. Anyone who knew the technology and who owned or controlled a bit of windy land could build a grain mill or a water pump. Previously, mechanical power had been a monopoly of those who controlled water power—generally the major land owners. This group did not react well to windpower, and their opposition to a threat to their monopoly of mechanical power resulted in the regulation or outlawing of windpower in many areas.

On the other hand, widespread use of windpower occurred in Holland and other low-lying areas which were without substantial water power. It was this same lack of water power which gave rise to a growth of decentralized windpower in the late nineteenth-century in the United States. As the western territory was opened to settlement farmers and ranchers turned increasingly to the windmill to pump water out of deep wells for livestock. When the transcontinental railroad was developed, water pumped by windmills provided the steam.

Wind energy systems worked very well in the American West and many were modified and developed to power electrical generators in the early twentieth century. The use of wind-generated electricity declined rapidly however with the development of the Rural Electrification Administration which created a vast network of power lines to rural areas. These power lines carried electrical current generated in distant power plants fueled by coal. Decentralized wind-generated electrical power gave way to centralized power generators which could be controlled by large economic interests and a new, vast market for commodity energy was developed.[11]

The history of wind energy in the world demonstrates the viability, effectiveness and independence of this form of solar energy. Wind-generated power is clean, efficient and available in many areas of the world. Except for some notable exceptions such as the large-scale use of windmills to drain low-lying land in Holland, wind energy has been used in a decentralized fashion. Such a use is still realistic and effective. It has been made considerably more effective by the development of systems for storage and use of energy in hydrogen. Onsite production of hydrogen by electrolysis and the use of the resulting hydrogen to produce electricity in the future in fuel cells could provide a major inducement for the use of decentralized wind energy.

While history offers few examples of centralized energy systems based on wind-generated energy, certain situations have made such centralization a likely possibility. Wind-generated electricity, similar to other forms of solar energy, is nonpolluting and totally renewable. In the plains states and coastal areas of the United States wind is a dependable and abundant resource. We have seen that a farm of about 5000 wind generators using present-day technology could produce 1 gigiwatt of power, or the amount of energy produced by a large conventional power plant. The area taken up by these generators would be substantial compared with the area at the site of the conventional power plant; however, the wind generators would require no coal shipped by rail from mines and would not cause air or water pollution. Wind-generated electricity could be converted into hydrogen in large electrolysis plants and then piped to distant points of use. This energy could then be reconverted into electricity in fuel cells or burned according to the need of the user.

The model solar-hydrogen economy which we are considering projects the gradual implementation of wind-generated energy as an essential aspect of the system. Both centralized and decentralized wind systems are projected. Wind generated electricity is envisioned to compose 0.8% of the total energy supply by T + 15, 3.6% by T + 45, and 33% in a mature solar-hydrogen economy.

Water Power

Water has long been a primary source of mechanical energy. Two basic methods are used to extract the energy directly from water. The first and most familiar uses the *motion* of water, as in the stationary water wheels in streams which provided energy for early grain mills and factories. The second uses the *heat* stored in the water. The amount of heat which large natural bodies of water contain varies with their depth. This variation is often manifested as a significant temperature differential which can be converted to usable energy. Ocean thermal energy conversion (OTEC) is based on this second way of extracting energy.

Kinetic energy of water can be stored for future use. It can also be manipulated much easier than wind. The most common means of using the kinetic energy of water is by forcing the moving water to turn a paddle wheel. First used in very early watermills, the same principle is applied to modern water-powered electrical generators. However, there are other ways of using the motion of water. Water moves in the oceans and lakes as waves and as tides, and both of these forms of motion can be used to generate electrical power.

In 1976 the installed hydroelectric capacity of the United States was about 13% of the total electrical energy generating capacity. In some countries such as Switzerland and Canada the hydroelectric capacity is much greater. In Switzerland for instance 87% of the installed capacity to produce electricity is based on water power. In Canada the figure is 60%.[12]

While the installed hydroelectric capacity of the United States at present represents a relatively small percentage of the total electrical energy capacity, this percentage could be increased significantly through the introduction of new forms of hydropower. Virtually the entire hydroelectric capacity of the world today is of one type. Damming rivers and use of water in these dams to make electricity has been the simplest and apparently the most economical way to harness the energy in water. The distance which the water can be made to fall as a result of a reservoir is called the *head* (Figure 3.6). The greater the height, the greater the head, and the more mechanical energy that can be used. High dams usually mean large water impoundments and involve vast areas of land. For some time there was a general professional and public assumption that all dams were beneficial. Not only could they provide vast amounts of electrical energy for important

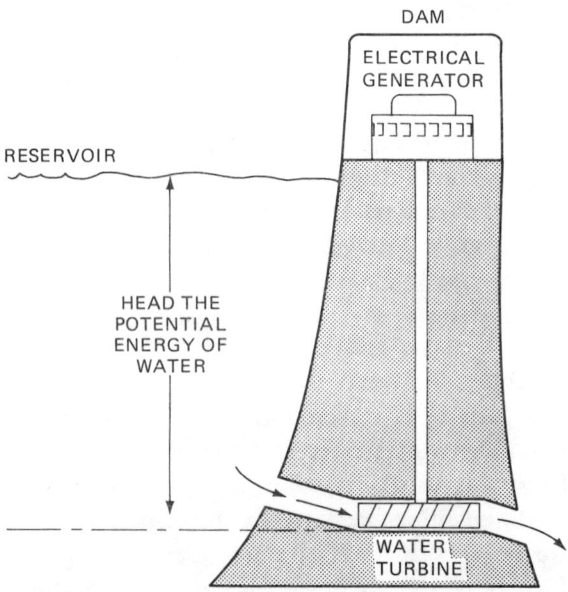

Figure 3.6 Principles of operation of conventional hydropower systems

public efforts such as rural electrification, but they also could be built to control flooding and improve the navigation of rivers. Flooding has been a recurrent problem in many areas of the United States. In the spring and fall when snow thaws and rains increase, rivers and streams overflow their channels and flood fields causing serious and expensive damage to homes and crops. If this overflow can be trapped by a dam and held in a reservoir until a drier season and then released, flooding could be reduced. In addition, dams combined with a judicious channeling of rivers could make rivers more dependable for navigation.

Unfortunately, the benefits of damming rivers are counterbalanced by serious environmental problems. High dams create vast changes in rivers and river-associated ecosystems. Habitats are changed and sometimes lost entirely. There is also now some doubt that dams actually reduce the danger of flooding. In fact, there is a possibility that over time dams may contribute to an increase in flooding.

Fortunately there are other, more environmentally benign methods for using the kinetic energy of water to produce electrical energy. Wave and tidal power are both important future sources of hydropower and both are abundantly available in the United States.

Wave power is in fact a form of wind power for it is the wind that causes the waves. Wind makes water molecules move in both horizontal and vertical directions so that a water molecule in a wave moves in a full circle (Figure 3.7).

Any effective wave power device must take advantage of this entire movement. The primary problem which must be overcome in wave-power technology is the natural dissipation of the kinetic energy of a wave. An object which bobs up and down in a series of waves is in motion, but all of the energy of that motion is returned to the water. In order to use the motion of a wave, that motion must be interrupted. One way is to convert the motion of the wave into a pumping action and use that action to create a flow of water. This method converts the circular motion of the wave into linear motion. A system designed by Stephen Slater converts the random forces of waves into pumping action through the use of a series of bouys or vanes which bob back and forth as a result of wave action and gravity. The Slater *ducks* as they are called have been model-tested with encouraging

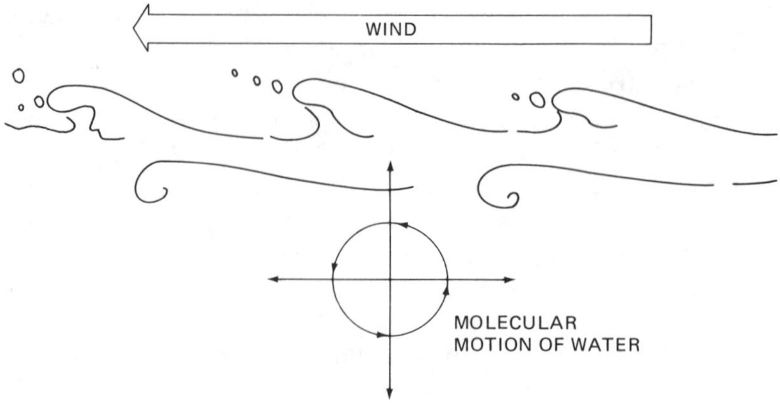

Figure 3.7 Wave energy and wind

results in Scotland.[13] The water which is caused to flow continuously through the system can be used to turn a turbine and run an electrical generator (Figure 3.8).

An earlier wave conversion design was created by Walter Bott in the Indian Ocean near the island of Mauritius. Waves break with regularity and intensity on coral reefs off the coast of the island. The original design used this wave force to run pumps that moved water into a reservoir with a head higher than the level of the coast. This water was then used to turn an electrical generator. Modifications of this design use the wave power to directly drive a generator or turbine.

Another system, designed by Sir Christopher Cockerell, uses the oscillation of waves to operate a series of pistons on hinged rafts. The

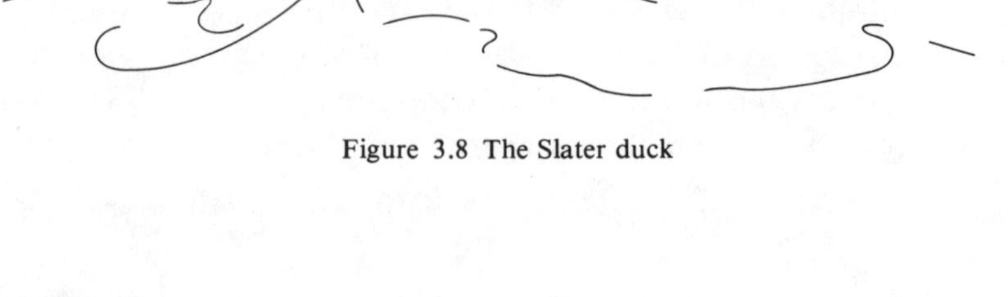

Figure 3.8 The Slater duck

rafts float on the surface and bend in opposition to each other with the action of the waves.

While the raft system depends on both the horizontal and vertical motion of waves, other systems use only the vertical oscillation of the water which results from wave action. Some of these designs such as the Japanese Masuda buoy use this vertical oscillation to create air pressure which drives a turbine. Others such as the Ocean Swell and Wave Energy Converter use vertical wave oscillation to move a conductor in a magnetic field to produce usable electrical energy.[14]

The conversion of tidal motion into energy is in principle based on the same physical actions as wave conversion. As with waves there is both horizonal and vertical motion to be used. However, tidal water molecules actually move in a vertical direction whereas water molecules in waves oscillate in a circular motion. Tidal water flows toward and away from shore and can be used to operate air and water pumps as well as to create reservoirs to store water at a higher level to run electrical generators.

The energy in the tides is a combination of solar and lunar energy. It is solar because the oceans depend for their existence on the sun, but tides exist because of the gravitational effects of both the moon and the sun. Vast amounts of water are moved regularly by the moon's gravity. This movement is manifested by the change in coastal water levels which can never be predicted exactly but can be estimated with a great deal of accuracy.

Considerable research has been devoted to tidal power and a number of potential systems have been conceptualized. It has been estimated that conventional tidal plants could be constructed at a capital investment similar to nuclear plants using 1976 dollar costs.[15] As nuclear power becomes more expensive tidal energy should become more financially attractive.

Wave and tide power have the advantage of a wide variety of conversion technologies which may be developed to take advantage of unique local conditions. Where waves are a result of steady and predictable wind patterns they tend to be regular and dependable. Under these conditions systems like Cockerell's rafts are a good choice. When waves vary in intensity, but have great vertical oscillation, systems like the Masuda buoys have a clear application. This variety of potential conversion technologies make wave and tide power an important future energy conversion process in the mixed centralized aspect of a solar-hydrogen economy. Wave energy systems could be built in comparatively small units which could be moved from area to area on the basis of energy demand. Such flexibility is extremely attractive in view of possible local changes in demand during the staging of a solar hydrogen economy. Tidal power also lends itself to permanent installation as a result of fairly constant local conditions. Both wave and tidal power should be important sources of water-generated electrical energy in all stages of the solar-hydrogen economy.

Ocean thermal energy conversion (OTEC) is another approach to the use of solar energy in water. The oceans of the world, particularly in the tropics, have very great *thermal gradients* between the warm surface water and the deep, cold water of the abyss. The difference in temperature between the surface water and the deep water is a relatively permanent condition. Water in the deep ocean does not move rapidly and there are very few current systems which return deep water to the surface. The heat energy of sunlight is trapped near the ocean's surface, and since there is little mixing of deep and surface waters a warm layer of water remains over a cold ocean. This temperature differential can be used to generate usable energy in a number of ways, all of which involve the flow of heat energy from the warm surface to the cold depths.

The system for energy conversion which has attracted most interest thus far is based on the *Rankine Thermodynamic Cycle*. This is a closed system in that it uses a working fluid not in direct contact with the ocean water to transfer heat energy. The most common working fluid is ammonia because it is readily available and it transfers from a liquid to a gas over a small range of temperatures. In a system using ammonia as a working fluid, the warm surface waters are used to evaporate liquid ammonia. The expansion of the ammonia gas is used to turn a turbine which runs an electrical generator. The gas is then cooled to a liquid in a condenser by cold water pumped from the deeper ocean. OTEC systems can be located onshore with ocean water piped in, or they can be located at sea. Both of these systems are under consideration (Figure 3.9).

While there are technical problems such as the growth of biological organisms on pipes, they seem to be surmountable. The reliability of

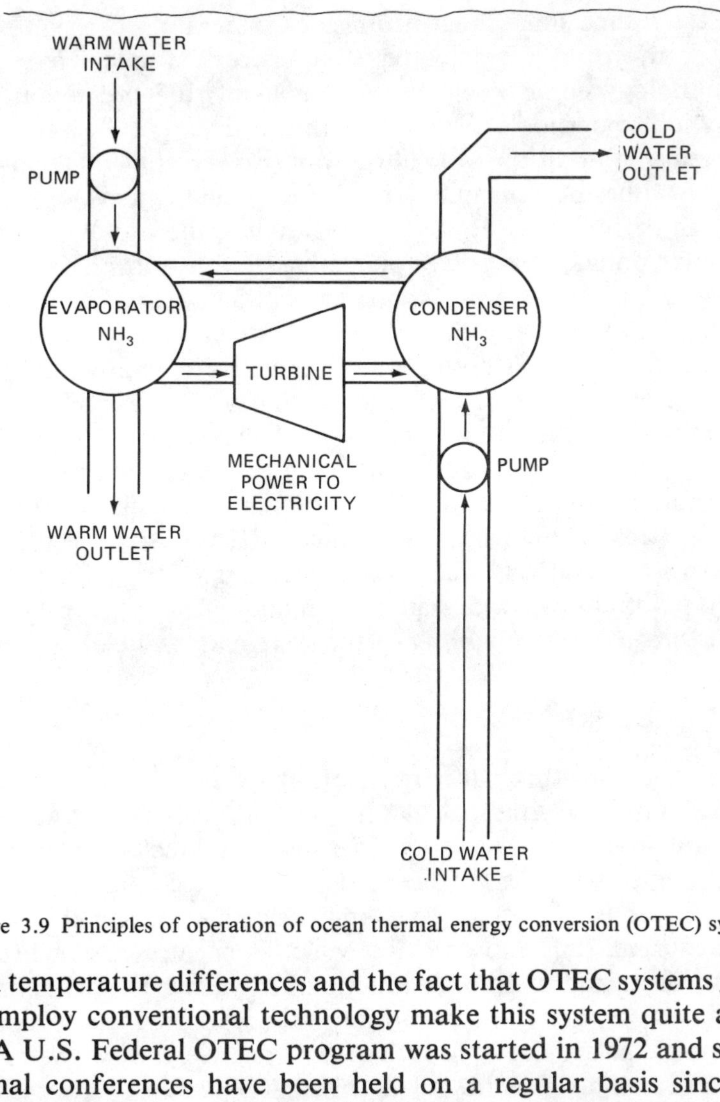

Figure 3.9 Principles of operation of ocean thermal energy conversion (OTEC) systems

ocean temperature differences and the fact that OTEC systems generally employ conventional technology make this system quite attractive. A U.S. Federal OTEC program was started in 1972 and several national conferences have been held on a regular basis since that time.[16] An example of a promising OTEC system is the Lockheed floating platform which would be located 30 miles from shore. About 950 feet long, this system would consist of four power modules and generate 240 MW of power. The system would have a design life span of 100 years for the platform and 35 years for the power modules.[17]

OTEC systems show great promise as integral parts of an overall water-based, solar-energy conversion process. Implementation of this technology could begin on a permanent, if incremental, basis before the last decade of the twentieth century.

There are in addition to OTEC, wave power, and tidal power, a number of other ocean technologies which could be developed in the next three or four decades. Among these are the use of ocean currents, solar ponds, and power plants based on osmosis. These technologies, while in the early stages, are potential renewable energy processes with few if any environmental side effects.[18]

Wind and water solar-conversion systems are generally site-specific and span a great range of technologies. They offer a number of advantages which make them an essential part of a solar-hydrogen economy. These systems are reliable, safe and environmentally benign. Another advantage of these conversion systems is their great diversity. Because there are so many different site-specific technologies with a relatively small cost in any single one, the system can undergo continuous, wide-ranging technological trial procedures without threat to the reliability of the system as a whole.

THERMAL CONVERSIONS

Categorizing solar-thermal conversion processes is difficult because there are so many technological applications of solar-generated heat. An accurate use of the term *solar-thermal* would include all wind and water systems as well as diffuse and concentrated solar-thermal device applications of both passive and active natures. In this section we will consider only those devices which concentrate solar heat for both the direct and indirect production of hydrogen. This approach should in no way detract from the importance of processes which employ diffused solar heat both actively and passively. Such systems include the familiar greenhouse; architectural designs of homes, offices, and factories which use sunlight as a stationary and integral part of the heating systems; and those designs which store and circulate solar heat. All of these systems are extremely important in a solar-hydrogen economy. They are generally decentralized, site-specific systems which reduce energy demands on centralized production systems. As such they represent a significant aspect of the overall energy production systems. While none of these systems is directly involved in the production of hydrogen they all reduce the demand for other forms of energy and thus contribute to the overall energy production system.

The design of devices to create intense heat from sunlight has long attracted human imagination. An early legend tells of the Greeks destroying an ememy fleet by focusing sunlight, reflected from mirrors, on the sails of invading ships. Solar-concentrating systems based on this same principle are now a reality.

Three classes of solar-thermal concentration systems have been identified by the U.S. Department of Energy as having the best economic potential for the United States.

The largest devices are found in a class called *central receiver systems.* These devices consist of a group of mirrors called tracking heliostats, which move so that they constantly reflect the maximum amount of sunlight. The reflected light is focused on a receiver point, usually on a tower. The temperature generated at the receiver point can be above 1500°F for gas-cooled systems. Heat produced in these systems can be used to generate steam to run electrical generators, or it can be used in the direct thermal splitting of water to produce hydrogen and oxygen (Figure 3.10).

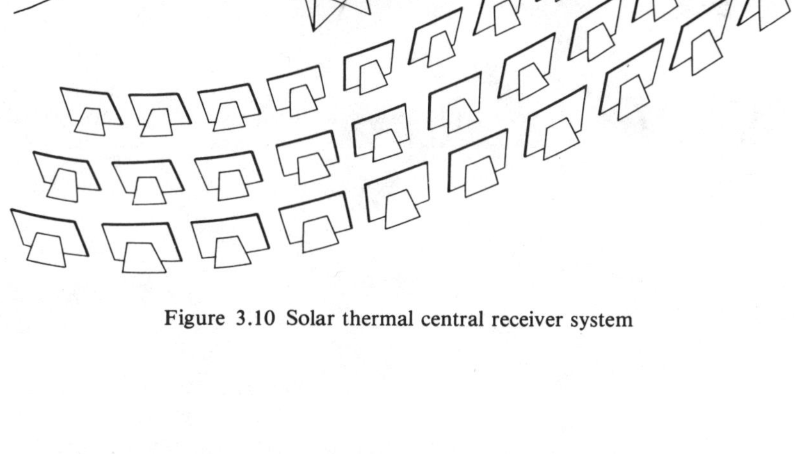

Figure 3.10 Solar thermal central receiver system

The other two classes of devices use shaped reflectors to focus light, either at a particular point—*point-focusing distributed receivers*—or along a pipe—*linear-focusing distributed receivers.*

Point-focusing devices are *parabolic dishes* which focus light toward one single receiver suspended in the dish. The dish tracks the sun to allow for continuous focusing. Parabolic dish devices currently operate to generate temperatures up to 750°F and they are expected to operate as high as 1500° in the near future (Figure 3.11).

Linear-focusing devices either focus light along a stationary pipe using parabolic troughs or along a moving pipe using hemispherical bowls. Parabolic trough systems may reach 600°f, whereas hemispherical bowl systems may operate as high as 1000°f (Figure 3.12).

All of these classes of devices fall into two categories based on the geometry of the reflector. Two-dimensional reflectors are trough-like devices such as half cylinders and parabolic troughs. Three-dimensional reflectors are bowls and dishes. Central receiver systems can be either two-dimensional or three-dimensional systems depending on the configuration of the individual reflectors.

Assuming no absorption of photon energy, *concentration of sunlight* (C) is by definition the ratio of the area upon which the light

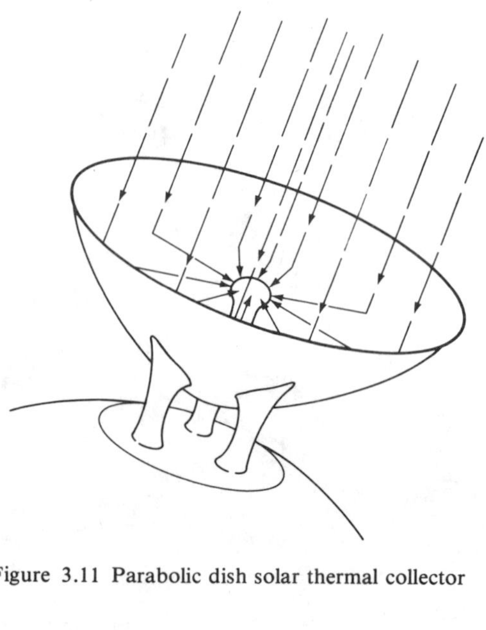

Figure 3.11 Parabolic dish solar thermal collector

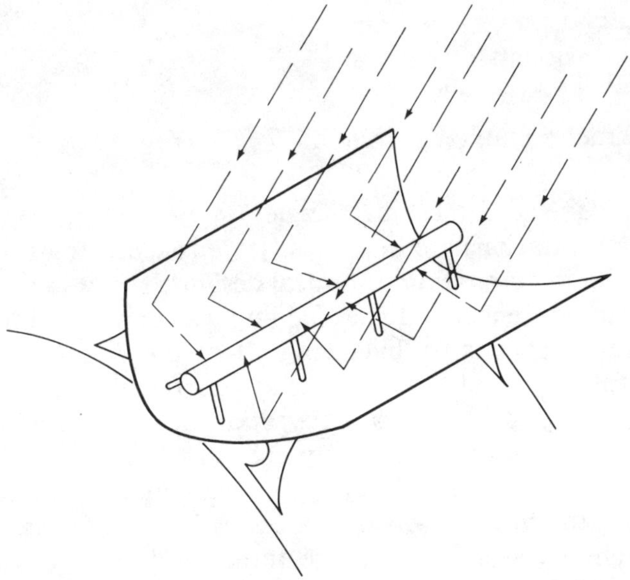

Figure 3.12 Parabolic through solar thermal collector

originally impinges—the *aperture area* (A^a)—and the area which receives the reflected light—the *receiver area* (A^r).

$$C = \frac{A^a}{A^r}$$

The effectiveness aperture area is to some extent determined by the shape and tilt of the collector. The shape and tilt determines the *angular range* which will accept the optimum amount of sunlight without movement of the collector. This range of angles is called the *acceptance angle* $2_{\emptyset c}$.

Thermodynamic limits of concentration are reached by concentrators which take advantage of the total *angular radius* of the sun, (0.25°). Devices which approach the thermodynamic limits of concentration are called *compound parabolic concentrators.* Such devices have been in use at least since the mid-nineteen-sixties.

The ideal concentration for these devices can be determined by the following formula:

$$C \text{ ideal (two dimensional)} = \frac{1}{\sin \varnothing_c}$$

$$C \text{ ideal (three dimensional)} = \frac{1}{\sin^2 \varnothing_c}$$

Where $\varnothing_c$ is the acceptance half-angle.

Ideal concentrations are never really approached for a number of reasons including limitation of optical designs, errors in tracking, atmospheric interference, and material imperfections. Optimal operating efficiency depends on balancing all of these conditions in the design of solar-thermal systems suited to particular areas.[20]

Concentrating solar-thermal conversion processes can be important in the solar-hydrogen economy in both the short and long term. The U.S. Department of Energy projects the application of concentrating solar-thermal conversion processes in three levels of market size through the year 2000. These three levels project the use of trough, bowl, dish and central receiver systems to provide electrical energy and industrial process heat in increasing amounts and reducing costs through the turn of the century. The DOE cost goals for the year 2000 are 50–60 mills/KWH for electricity and 4–5 dollars/MBtu for heat. While these cost goals are under review and should not be considered firm at this point, they do indicate the serious potential for solar thermal conversion processes.[21]

In an emerging solar-hydrogen economy, solar-thermal energy is stored and transported as hydrogen. In a mature solar-hydrogen economy the conversion of solar-thermal energy to hydrogen will take place through the heat splitting of water, a process which will be discussed in the next section.

During the implementing stages of the economy solar-thermal energy can be used directly as heat in industrial process or converted to electrical energy in generators. The electricity produced can be used as electrical current near the site of generation or converted to hydrogen through electrolysis. The hydrogen produced can then be piped to distant locations and stored for future conversion to heat or electricity near the site of use.

Both diffused and concentrated solar-thermal energy are important elements in a solar-thermal energy process. Diffused energy can be employed to reduce commercial and residential dependence on centralized energy sources. Concentrating solar-thermal devices on the other hand are an increasingly attractive alternative to conventional power-generating systems, and in addition they are potentially a direct means of producing hydrogen through the process of thermal water splitting.

NOTES

1. R. W. Foster et al., *Solar Hydrogen Systems for the 1985–2000 Time Frame.* U.S. Department of Energy, June 1980.
2. Isaac Trachtenberg, *The Texas Instruments Solar Energy System,* Central Research Laboratories, Texas Instruments Incorporated, Dallas, Texas, presented in Houston, Texas, June 1–3, 1982.
3. E. L. (Pete) Johnson. *Development of a Solar Energy System: A Special Technical Summary Report.* Texas Instruments Incorporated, for U.S. Department of Energy, DE-AC01-79ER 10000, April 3, 1981.
4. An Associated Press coverage of a news release from Texas A & M on October 8, 1982. Also Thomas H. Maugh II, "Two New Routes to Solar-Hydrogen", *Science* 218, November 5, 1982.
5. Arthur H. Fierster, *Amorphous Silicon Solar Panels: Cost, Production, and Performance Projections* (Princeton: RCA Laboratories n.d.)
6. A. Catalano et al., *Attainment of 10% Conversion Efficiency in Amorphous Silicon Solar Cells* (Princeton RCA Laboratories, n.d.)
7. Fierster.
8. Paul Danish. "Low Cost Silicon and PV Devices Could Result from SERI Process," *Solar Times,* December 1981.
9. Jerrold H. Krenz. *Energy Conversion and Utilization* (New York: Allyn and Bacon, Inc., 1976), p. 283.
10. William L. Dickinson and Paul N. Cheremisinoff. *Solar Energy Technology Handbook* (New York: Marcel Dekker, Inc., 1980).
11. Ibid.
12. Charles Simeons. *Hydro-Power, the Use of Water as an Alternative Source of Energy* (New York: Pergamon Press, 1980).
13. Ibid.
14. David D. Woodbridge, "Source and Potential Uses of Wave Energy" in *Alternative Energy Sources II,* ed. T. Nejat Veziroglu (Washington: Hemisphere Publishing Corporation, 1981).
15. Alexander M. Gorlov, "Some New Conceptions in the Approach to Harnessing Tidal Energy," in *Alternative Energy Sources II,* ed. T. Nejat Veziroglu (Washington: Hemisphere Publishing Corportion, 1981).
16. *Ocean Systems Program Summary,* U.S. Department of Energy, DOE/ET-0083, Washington, D.C., February 1979.

17. Leon W. Zelby, "A Perspective on OTEC Plants," in *Alternative Energy Sources II,* ed. T. Nejat Veziroglu (New York: Hemisphere Publishing Corporation, 1981).
18. *Solar, Geothermal Electric and Storage Systems,* Program Summary Document, U. S. Department of Energy, DOE/ET-0102, July 1979.
19. *Solar Energy,* Program Summary Document, FY 1981, U. S. Department of Energy, DOE/CS-0050, Washington, D.C., August 1980.
20. Ari Rable, "Concentration Collectors," in *Solar Energy Technology Handbook,* ed. William C. Dickinson and Paul W. Cherimisinoff (New York: Marcel Deckker, Inc., 1980).
21. *Solar Energy,* Program Summary, August 1980.

REFERENCES

Charles, H. K., Jr. and A. P. Ariotedjas. "Review of Amorphous and Polycrystalline Thin Film Silicon Solar Cell Performance Parameters," *Solar Energy* 24.

Danish, Paul. "Low Cost Silicon and PV Devices Could Result from SERI Process." *Solar Times.* December 1981.

Department of Energy. *Photovoltaic Power Systems Market Identification and Analysis.* Washington, U. S. Government Printing Office, 1978.

Dixon, A. F. and J. D. Leslie, eds. *Solar Energy Conversion.* Toronto: Pergammon Press, 1978.

European Cooperation Space Environment Committee. *Solar Cells.* London: Gordon and Brouch Science Publishers, 1971.

Ehrenreich, Henry and John H. Martin. "Solar Photovoltaic Energy." *Physics Today* 32. September 1979, pp. 25–32.

Foster, R. W. et al. *Solar Hydrogen Systems for the 1985–2000 Time Frame.* U. S. Department of Energy, June 1980.

Gorlov, Alexander M. "Some New Conceptions in the Approach to Harnessing Tidal Energy" in *Alternative Energy Sources II,* ed. T. Nejat Veziroglu. Washington: Hemisphere Publishing Corporation, 1981.

Hautala, Richard R. et al. *Solar Energy Chemical Conversion and Storage.* Clifton, NJ: The Humona Press, 1979.

Johnson, Dale G. et al. *Hydrogen From Existing Falling-Water Sites: An Initial Feasibility Assessment.* Chicago: Institute of Gas Technology, 1978.

Kelly, Henry. "Photovoltaic Power Systems: A Tour Through the Alternatives." *Science* 199, February 10, 1978.

Maycock, Paul D. and Edward N. Stirewalt. *Photovoltaics.* Andover, MA: Brick House Publishing Company, 1981.

Merrigan, Joseph A. *Sunlight to Electricity.* Cambridge, MA: The MIT Press, 1975.

Ocean Systems Program Summary. U. S. Department of Energy. DOE/ET-0083. Washington, D. C., February 1979.

Pulfrey, David L., Ph.D. *Photovoltaic Power Conversion.* New York: Van Nostrand Reinhold Company, 1978.

Rable, Ari. "Concentration Collectors" in *Solar Energy Technology Handbook,* eds. William C. Dickinson and Paul W. Cherimisinoff. New York: Marcel Dekker, 1980.

San Pietro, Anthony. *Biochemical and Photosynthetic Aspects of Energy Production.* New York: Academic Press, 1980.

Simeons, Charles. *Hydro-Power, the Use of Water as an Alternative Source of Energy.* New York: Pergamon Press, 1980.

Singh, R. and J. D. Leslie. "Economic Requirements for New Materials for Solar Photovoltaic Cells." *Solar Energy* 24.

Sittig, Marshall. *Solar Cells for Photovoltaic Generation of Electricity.* Park Ridge: Noyes Data Corporation, 1979.

Solar Energy. Program Summary Document. FY 1981. U. S. Department of Energy. DOE/CS-0050. Washington, D.C., August 1980.

Solar. Geothermal Electric and Storage Systems. Program Summary Document. U. S. Department of Energy. DOE/ET-0102, July 1979.

Solarex Corporation. *Making and Using Electricity from the Sun.* Blue Ridge Summit, PA: Tab Books, 1979.

Stewart, John W. *How to Make Your Own Solar Electricity.* Blue Ridge Summit, PA: Tab Books, 1979.

Technology for Solar Energy Utilization. New York: United Nations, 1978.

Veziroglu, T. Nejat. *Àlternative Energy Sources II, Vol. 3, Solar Energy 3.* Washington: Hemisphere Publishing Corporation, 1981.

Woodbridge, David D. "Source of Potential Uses of Wave Energy" in *Alternative Energy Sources II.* ed. T. Nejat Veziroglu. Washington: Hemisphere Publishing Corporation, 1981.

Zelby, Leon W. "A Perspective on OTEC Plants" in *Alternative Energy Sources II.* ed. T. Nejat Veziroglu. New York: Hemisphere Publishing Corporation, 1981.

4. ENERGY TO HYDROGEN

When solar-produced energy is converted into hydrogen it can be stored in the hydrogen molecule for an indefinite period. It can also be transported from a place of conversion to any point of use.

Three basic processes can be used to make this conversion. The first can be broadly categorized as *biochemical* and further reduced into two types. One involves the conversion of biologically produced materials such as coal, natural gas or even wood to hydrogen. In this type of process hydrogen is released from the chemical structure of the biological material, or from its bonding with oxygen in water, as a result of heat from the combustion of a biological material. The other type of biochemical process, often called the *photobiotic* conversion, involves photosynthesis or photosynthesis mimicking processes to produce hydrogen directly.

The second category includes all of the *thermal* conversions which are solar related. These processes involve the heat dissociation of a hydrogen-bearing substance, primarily water.

The final category can be called *electrochemical.* In this process water or some other hydrogen-bearing compound is dissociated by electricity.

Selected aspects of each of these categories of conversion processes will be discussed with regard to basic principles, technology, research potentials, and applicability to the model solar-hydrogen economy.

BIOCHEMICAL CONVERSIONS

Coal

Hydrogen is commonly produced from natural gas. However, the production of hydrogen from coal is considered the most interesting biochemical process for the near-term staging of a solar-hydrogen economy. This is so not because it is cost effective at the present time but rather because of its potential. The availability and abundance of coal, the rising cost of gasoline and natural gas, and the market externalities associated with the uncontrolled combustion of hydrocarbon fuels indicate the increasing importance of coal gasification to hydrogen.

On the other hand, the difficulties associated with bringing coal back to prominence as a major fuel are well recognized. It will require considerable encouragement from the public sector. It is also clear that, without a major planning decision, production of hydrogen from coal will probably not be given much emphasis.[1] However, the gasification of coal to hydrogen has some important middle-range benefits which are probably essential to the implementation of a solar-hydrogen economy. Besides being abundant and relatively inexpensive, coal is familiar as a raw material; and coal gasification itself, which may seem new to this generation, is in fact an old technology. Gasification to hydrogen is a new step in the process, but a small step in terms of technology implementation. This technology is seen as a way of introducing a substantial increase in the use of hydrogen as a domestic fuel in the early and middle stages of the model. But to better understand this technology, it might first be helpful to review some characteristics of its basis, coal.

The familiar substance, coal, which has dominated so much of the history of industrial society, is a very complex material. It is not a consistent substance, but rather consists of a range of similar mineral mixtures formed out of a combination of biochemical and geological processes. Peat deposits, composed of plant remains and some inorganic materials, have been compressed by sediment and geological movements of the earth over a long period of time. This process has resulted in various mixtures of organic compounds and inorganic materials such as sulfur.

It has been a great challenge and a considerable frustration for engineers to classify coal. The results of this effort are as imprecise as the material itself. In general, coal is classified according to rank, type and grade. *Rank* is a reflection of the degree of geological alteration, or metamorphosis, of the original material. *Type* is a measure of the diversity of materials (heterogeneity) which composed the original mix. *Grade* reflects the purity of the coal (moisture, mineral matter, and organic substances which are not fully combustible are considered to be impurities).

Rank is the most common way of categorizing coal. As a reflection of the effects of geological alteration, it is a measure of the carbon and oxygen content, and the general heating potential of coal. There are many ways of ranking coal, including an international system. The system most commonly used in the United States has been developed by the American Society for Testing Materials (ASTM). The three concepts routinely used by the ASTM classification system are: *fixed carbon* content which is the amount of coal not combusted when a sample is heated to 950°C for seven minutes, *heating value* which is the heat given off when a sample is completely oxidized, and *agglomeration* which is the carbonized residue from a sample burned under controlled conditions.

These three concepts are applied to define four basic ranks of coal: *anthracite,* which has the highest carbon content, the highest heating value and is more agglomerating in character; *lignitic,* which has the lowest amount of fixed carbon, the lowest heating value and which is also nonagglomerating in character, and *bituminous* and *sub-bituminous* which fall between these two ranks and have some agglomerating characteristics.

While the existence of coal was recognized as far back as the twelfth century it did not come into general use until an energy crisis developed in Europe in the seventeenth century as a result of deforestation. In the early eighteenth century a process was developed for producing coke, a high carbon material, from coal for use in making steel. The use of coal in industry increased from this period and spread to other aspects of the energy economy until it had fully eclipsed wood as the dominant fuel.

There are several energy conversion processes involving coal; the most common of these, and the easiest to understand, is combustion. In this process coal is simply oxidized rapidly and heat is released. This process of combustion was the foundation of the carbon-combustion economy of the industrial era. Between the world wars, as a result of a range of political, economic and technical developments, the combustion of coal began to give way to the combustion of oil as the world's major fuel. World War II completed the eclipse of coal by oil. However, the use of coal still remains an essential element of the carbon-combustion economy, particularly in the generation of electricity.

There is however another important coal conversion process which predates combustion as a major coal use. This process is called *pyrolysis* and is essentially the heating of a substance in a non-oxidizing environment. This process causes coal to change its mixture characteristics. The first use of the process was to create coke to replace the charcoal used in making steel. The same process is used to make coal into a gas and or a liquid. Coal gasification is of particular importance in the early stages of a solar-hydrogen economy because it provides a transition from a familiar technology to a new pattern of energy distribution.

The use of gas made from coal was quite prevalent prior to the availability of cheap natural gas, which came into use with the development of the oil industry. During the 1920s in the United States there were between 10,000 and 20,000 coal gasifiers. The gas produced in gasifiers is generally of three types. Depending on the energy content it can be low Btu, medium Btu or high Btu. Though hydrogen is the main energy producer in each of these types it is generally not produced in a pure form. The technology for such production is relatively recent and depends on the use of metal hydrides and other separating techniques.

The chemical process involved in the production of pure hydrogen from coal parallels traditional coal gasification processes until the last stages. The following is a simplified description of the process of coal gasification to hydrogen (Figure 4.1):

1. Coal is dried and heated to remove water.
2. Coal is further heated to drive out easily volatilized hydrocarbon and water trapped in submicroscopic pores.
3. The remaining coal material is exposed to air and steam at the temperature of ignition. The following reactions take place:

A. $C + O_2 \rightarrow CO_2$

$$2H_2 \rightarrow 2H_2O$$

These two reactions are combustion processes necessary to provide heat for future reactions.

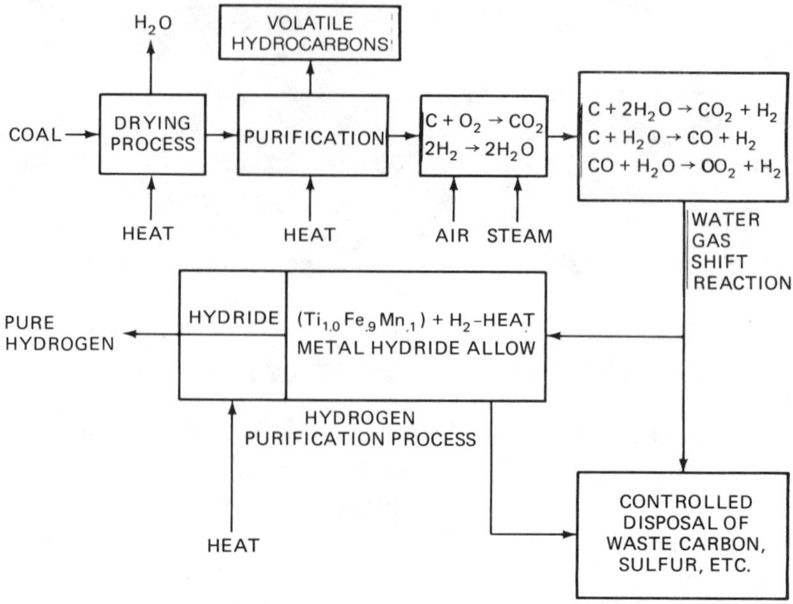

Figure 4.1 Goal gasification to hydrogen

B. Hydrogen is formed in a number of reactions sometimes proceeding simultaneously:

1. $C + 2H_2O \rightarrow CO_2 + H_2$

This reaction approaches equilibrium but not below 980°F.

2. $C + H_2O \rightarrow CO + H_2$

The above reaction is endothermic but never reaches equilibrium.

3. $CO + H_2O \rightarrow CO_2 + H_2$

The above reaction is called the *water gas shift reaction,* and it approaches equilibrium.

C. Hydrogen, if not removed, combines with carbon to make methane in a hydrogasification process. This process is basic to the medium Btu gasification processes:

$C + 2H_2 \rightarrow CH_4$
$CO + 3H_2 \rightarrow CH_4 + H_2O$
$CO_2 + 4H_2 \rightarrow CH_4 + 2H_2O$

Since methane (CH_4) is the desired end product in traditional gasification processes hydrogen is added from other sources to combine with carbon in this process.

D. Since hydrogen and not methane is the important product in coal gasification to hydrogen, the problems at this point are the removal of pure hydrogen and the removal of carbon- and sulfur-related wastes for disposal.

1. Carbon-related pollutants can be removed in a number of ways. Reactions based on potassium carbonate are among these: $CO_2 + H_2O + K_2CO_3 \rightarrow 2KHCO_3$ (potassium bicarbonate), H_2S (formed when sulfur impurities combine with hydrogen) + $K_2CO_3 \rightarrow K_2S$ (potassium sulfide) + H_2CO_3 (carbonic acid).

2. Until recently no convenient and acceptable method was known for removing pure hydrogen from this process. The discovery of the practical use of metal hydrides has provided that method. Hydrogen, when cooled and exposed to certain metal alloys, is absorbed into the molecular structures of these alloys in a pure form. This absorption is so complete that more hydrogen molecules can be absorbed in a metal hydride than can exist in an equal volume of pure liquid hydrogen. Hydrogen of research grade purity can be extracted under pressure from the metal hydride by the addition of residual heat from the gasification process.

metal hydride

$$\left[\begin{array}{l}\text{Iron titanium alloy}\\ (Ti_{1.0}Fe_{.9}MN_{.1})\text{ or}\\ \text{other alloy}\end{array}\right] + H_2\text{ -}\overset{\uparrow}{\text{(Heat)}} \rightarrow \begin{array}{l}\text{Metal}\\ \text{Hydride}\\ \text{Storage of}\\ \text{Hydrogen}\end{array}$$

The process described above is technically available now and there are presently several commercially available systems for the early stages of the process. Early stage systems are considered in three major categories of coal gasification: fixed or moving bed, entrained flow, and fluidized bed.

In the fixed- or moving-bed system, coal is put in at the top and moves down through the gasifier; steam and air are introduced in the bottom of the coal bed; carbon reaction with oxygen takes place at the bottom; and the water gas shift takes place throughout the coal bed. A serious problem, associated with this system, is the tendency of the coal to cake in the bed. Early systems of this type were produced by the Lurgi corporation and there are several other corporations such as Wellman-Galusha and General Electric involved in the development of modern, low Btu gasifiers of this type.

Entrained flow systems use the mixture of air and steam to move coal particles through the gasifier. There are several companies involved in developing low Btu gasifiers of this type. The Shell Oil Company using a Koppers-Totzek pressurized system, the Babcock and Wilcox Company, Texaco and Foster Wheeler are among these.

Fluidized bed systems were introduced in Germany prior to World War II. These systems involve the introduction of preheated coal bed into a gasifier. The only commercially available fluidized-bed system at present is the Winhler process which was originally introduced in Germany and later modified to use oxygen and steam.

All of these systems are capable of modification to produce hydrogen. One design for a moderately sized plant appropriate to the production of hydrogen for use by an industry or a small municipality has been proposed by L. D. Hadder of the Billings Energy Corporation.[2] This system, is a fixed-bed process which possesses great stability and flexibility, though like all fixed- and moving-bed systems its output is relatively lower than the less developed fluidized-bed systems.

While most of the hardware necessary for the introduction of coal gasification to hydrogen systems is now available, developmental and research work needs to be done on the pollutant and hydrogen removal aspects of the system. These efforts, combined with whole system design development, could result in an operationally tested and dependable plant by 1985.

The model solar-hydrogen economy projects the use of coal to increase to 52% of the entire energy use profile by the year T + 15. Of this 52% almost one-half of this amount, or a total of 20% of energy produced, will be supplied by hydrogen produced by coal gasification. By the year T + 30 all coal use in the United States would be in coal gasification processes. While the years between T + 30 and T + 95 will see the shift of all coal use from energy-producing process to chemical-producing process, the use of coal in a coal gasification system is an essential aspect of the staging of a solar-hydrogen economy.*

Photosynthesis and Photosynthesis Mimicking. Photobiotics is the most challenging, the most dangerous, and perhaps the most promising aspect of the solar-hydrogen economy. In simple terms photobiotics is the production of hydrogen through photosynthesis (in vivo) or a duplication of some apsects of that life process (in vitro). This method of production is still in the early research stages and is not projected to be a major apsect of the model until well into the twenty-first century. By then humankind may have addressed, through enlightened, careful choices, the crucial issues of microbiological engineering. The extent and the kind of manipulation which is imposed on the micro aspect of the terrestrial life system may well determine the next stages of evolution regardless of the type of energy economy.

Photosynthesis is the process by which certain organisms, such as green plants, dissociate the water molecule and use hydrogen in combination with carbon dioxide to produce carbohydrates. In this process photons of light supply the necessary energy for molecule build-

---

* For an interesting and important coverage of the principles and technologies of the production of hydrogen from fossil fuels see K. E. Cox, and K. D. Williamson, Jr., eds. *Hydrogen: Its Technology and Implications,* Volume I, Production Technology, (Cleveland: CRC Press Inc., 1977), pp. 61–115.

ing. It is possible in some cases to interrupt this process of molecule building before the hydrogen has combined with carbon dioxide to form carbohydrates. Approaches which are conceived to make the interruptions are called *in vivo* systems because they involve living cells. Methods for interrupting the process of photosynthesis in microorganisms are still very much in the research stages. Research efforts in these methods are all more or less concerned with the activities of two groups of enzymes involved in the process of hydrogen use and evolution. *Hydrogenase* is a group of iron-sulfur proteins which are involved in the metabolism of the hydrogen molecules according to the following basic formula:

$$2H^+ = 2e^- + H_2$$

Hydrogenase enzymes were most useful in the earliest stages of the development of atmosphere on this planet. These enzymes allowed organisms to conduct useful energy transfers employing photons and hydrogen. Several kinds of organisms still employ this process. Most photosynthetic bacteria have hydrogenase and it is also found in some protozoans, algae, higher plants and animals.[3]

Nitrogenase are enzymes composed of protein complexes of iron, sulfur and molybdenum. These enzymes are primarily involved in the fixation of nitrogen but they can also catalyze hydrogen.

Several different approaches have been taken toward the development of photobiological hydrogen production systems. One such approach has been to grow organic material with characteristically higher hydrogenase activity. Photosynthetic bacteria such as *Rhodospirillium rubrum* have been cited as offering the best potential in the short run.[4] Green and red algae (sometimes considered a bacteria) have also been the subject of considerable research. In the long run however genetic engineering and bioengineering of wild strains of organic material is the most likely course of research.

In genetic engineering a wild strain of an organism is selected and its genetic structure is manipulated through mutation or applications of recombinant DNA in order to enhance the photosynthetic metabolism of hydrogen.

One of the major impediments to the photobiological production of hydrogen is oxygen which, among other things, inhibits the activity of hydrogenase. Oxygen can be physically removed from the process by removing it from the medium of growth, but this process of removal is prohibitively expensive. Another way of inhibiting oxygen activity is to separate the oxygen-producing mechanism of the organism from the hydrogen-producing mechanism. These processes can be separated either by space or by time. Organisms in which both the photosynthetic production of hydrogen and oxygen occur in the same cell must be manipulated so that these two processes happen at different times. This separation can be accomplished through bioengineering.

In vitro photosynthetic hydrogen-production systems range from non-cell processes closely resembling natural photosynthesis to completely non-vital systems which use metals to capture the energy of photons and use this energy to split water. The range of processes can be considered in two categories: non-cellular photosynthetic systems, and systems which attempt to duplicate aspects of the natural photosynthesis systems.

Non-cellular processes involve the use of the same photosynthetic chemicals which are used in living photosynthesis. These materials are separated out of one or more living organisms and used to produce hydrogen. One research direction involves the use of spinach chloroplasts linked together with hydrogenase or a similar but synthetic substance.[5] This process has been under examination since the early 1970s but it has not yet been demonstrated to produce hydrogen over a sustained period of time sufficient for practical application.

Another cell-free research possibility involves the use of the protein bacteriorhodopsin, called the *purple membrane,* which contains retinal. It has been suggested that isolated purple membrane, because of its stability, ease of preparation, and high level of photopotentians, could be used to generate hydrogen.[6]

Several research directions are examining the duplication of aspects of natural photosynthesis. The problems of duplicating photosynthesis involve capturing photons, essentially solved by the pigmentation of chlorophyll in green plants, and using the energy of the captured photons to split water into hydrogen and oxygen. The process of using photon energy also involves the establishment of a dependable, nonstatic, electrical field. The process of converting photon energy has been the subject of most of the research thus far.

One recent effort involves using a catalyst composed of cadmium sulfide doped with ruthenium dioxide immersed in water and hydrogen sulfide. So far, research efforts using this catalyst have experienced problems with side reactions and oxygen/hydrogen separation. However, this research is particularly interesting because it uses hydrogen sulfide, one of the most abundant and noxious of industrial wastes, as a medium.[7]

The photobiotic conversion of sunlight to hydrogen, and systems which seek to duplicate this process, are not included as essential aspects of the model solar-hydrogen economy because these conversions are in the very early stages of research development. They are however among the most important of the long-term research efforts. These conversion processes offer the long-range possibility of combining hydrogen fuel production with commercial and residential waste disposal in a clean and effective system. This potential combination of uses makes them an important research and development possibility in a mature solar-hydrogen economy.

Thermal Conversions

All of the conversion processes we have considered involve the dissociation of the water molecule into hydrogen and oxygen. We have observed that a common method of converting solar energy to hydrogen for storage and transportation is through the electrochemical dissociation of water. There are several ways of converting solar energy into electricity in order to accomplish electrolysis. Methods which we have considered essential to the solar-hydrogen economy involve the conversion of wind and water mechanical energy to electricity and the various photovoltaic processes. One process we have not considered is the traditional process of the combustion of solar-produced biomass to make heat in order to run steam turbines. The resulting alternating current must then be converted to direct current in order to dissociate water into hydrogen and oxygen. While this method may have some short-term application as a use for excess electricity, it uses so much energy in its various steps that it is clearly inefficient. This inefficiency is all the more intolerable when we realize that it is not necessary to convert heat into mechanical energy, then into electrical energy, in order to disassociate the water molecule. Heat alone, or in conjunction with various chemical reactions, will separate water into hydrogen and oxygen. The heat necessary for this process is available from sunlight through single- or multiple-stage processes depending on the temperature generated.

At temperatures above 1500°K, well within the range of a number of solar thermal systems mentioned in Chapter 3, water will split into hydrogen and oxygen. Essential problems with this process include the difficulties in separating the two products before they reach a lower temperature and recombine to form water, and the problems associated with sustained high-temperature operation.[8] However, there are several techniques which hold potential for accomplishing separation, including membrane diffusion and centrifugal separation. The problems associated with high-temperature operations are in the process of research and will undoubtedly require considerable system design development. While more operational research is needed in the area of separation, high-temperature solar-thermal dissociation of water has great potential.

High-temperature dissociation can be considered a single-stage process. Multistage processes of dissociation can employ much lower temperatures, in the range of 800°K. These processes generally involve either the oxidation of a suitable metal in water with an ensuing release of hydrogen followed by a heat reduction of the metal to release oxygen, or a metal hydride capture of hydrogen and a subsequent release of the element. Both of these process methods are completely reversible and involve the formation of no secondary compounds. The metals involved are theoretically 100% recyclable. Comparatively little effort has thus far been devoted to the development of commercial processes for the solar-thermal dissociation of water. However, because of its inherent potential efficiency as a process for the production of hydrogen on a large, centralized scale, it has been selected as a major component of a mature solar-hydrogen economy. The solar-hydrogen economy model envisions fully 15% of the national energy production to be supplied by this method. In order to accomplish this level of production several experimental demonstration units would have to be constructed in the T + 45 stage of the model. These units would be integrated into all three use corridors.[9]

ELECTROCHEMICAL CONVERSIONS

A fundamental element in the solar-hydrogen system is the electrochemical production of hydrogen. This element is a two-stage process involving first the conversion of sunlight into electricity and second the use of this electricity to dissociate the water molecule.

We have observed that the first stage of this process can be accomplished in a number of ways. The most common method is the conversion of sunlight into biomass. The heat from the combustion of biomass such as wood, coal, or oil to produce heat is used to make steam. Finally, the steam is used to turn an electrical generator. This multistep process involves an energy loss at every transition, yet it is at present the primary means of producing electricity.

Another means of producing electricity from sunlight is to use the mechanical energy, generated terrestrially by solar heat in the form of wind and water power, to run an electrical generator. This method of electricity production is more efficient than carbon combustion. An even more direct method of producing electricity from sunlight is the conversion of solar radiation into electricity through the photoelectric effect. We have observed that there are a number of ways of making that conversion under the label of photovoltaics.

Once electricity in the form of direct current is produced, it can be used to dissociate the water molecule in an electrocatalytic process. This process can be simply the immersion of two ends of a conductor in a salt solution or a complex combination of techniques to create a photoelectric chemical system which joins the photoelectrical production of electricity with electrolysis. Whether simple or complex these systems all involve water and electrical energy in a particularly intimate relationship.

Electricity can be considered a chemical process involving the fundamental aspects of atomic structure, primarily the electron in the outer orbitals of molecules. In earlier sections we considered the nature of electrical current. We noted that electrical current involved the communication of energy between molecules through the interaction of electrons in the outer electron orbitals. Some of these electrons receive enough energy to reverberate out of a stable relationship with the remainder of the molecules. If these electrons move far enough they leave behind an electron hole in the molecular structure. This hole can be filled either by the previous electron or another electron which has been excited into activity. Electrons in an excited state pass their energy of excitation to other electrons in a pattern of displacement and replacement as long as the source of their excitation remains.

Voltage is the application of an imbalance of electrons, *negative charges,* and electron holes, *positive charges,* across a particular space. This difference in potential creates an electrical field. The molecular instability which could be caused by the application of a particular voltage in a particular material is called *current.* Current itself is a function of both the voltage and the molecular nature of the material. The molecule's nature is expressed in a concept called *resistance* which is a measure of the reluctance of valance electrons to become excited enough to transfer energy. The relationship was expressed by Georg Simon Ohm in 1826.

$$I \text{ (current)} = \frac{E \text{ (voltage)}}{R \text{ (resistance)}}$$

If a materal is composed of molecules in which the outer electron orbitals are stable to the extent that they require a great amount of energy in order to become excited enough to move, the material has great resistance and is a nonconductor of electrical current. Paper and rubber are in this category. If a material is composed of molecules which have very excitable outer electron orbitals it then has low resistance and is a good conductor of electrical current. We have already discussed those materials known as semiconductors. These materials, such as silicon and germanium, are made of molecules containing outer electron orbitals which are reluctant to excite but which nevertheless can be freed with the application of a particular amount of external energy. This energy is the *breakdown potential* and when it is applied, the valance electron of particular semiconductors are excited out of the *valance band* into the *conduction band.* These bands are ranges of energy, called the *energy gap.*

When the energy applied to semiconductors is in the form of photons these semiconductors can be made to function as photovoltaic devices and electrical current can be produced. When energy is applied, electrical semiconductors can be used to control precisely and incremently the transmission of electrical energy. Such control is the basis of microelectronics and finds a familiar form in transistors.

Some molecules, called *ions,* carry a residual charge equal to a multiple of their electronic charge. Materials which dissociate into ions are called *electrolytes.* When an electrolyte is dissociated into its ions, it is capable of conducting electrical current as a result of the migration of its ions. Ohm's law holds for an electrolyte just as it does for any other conductor.

The *standard equivalent conductance* of ionic solutions or electrolytes has been calculated and the relative efficiency of electrical conduction is known for a range of substances. As might be expected the most conductive is the hydrogen ion. Common electrolytes include potassium chloride (KCl), hydrochloric acid (HCl) and sodium hydroxide (NaOH).

Electrolytes are often divided into two categories, *acidic* and *basic.* An acid may be generally defined as a chemical species with a desire for electrons. In other words an acid is a substance with more protons than electrons. This imbalance is manifest in the outer electron orbital. A base on the other hand is a substance that wants to yield electrons. The strength of a liquid electrolyte is determined by its degree of disassociation into ions when dissolved in water. The more fully it dissociates the greater its strength. Such an electrolyte is considered strong if it is fully dissociated into its ions when it is dissolved in water.

When an acid such as hydrocloric acid (HCl) is dissolved in water it acts as though it were actually two different molecules. Through a process of *deprotonation* called *hydrolysis* the acid forms the *hydronium* ion. In this case water acts as a base and the hydronium ion H_3O^+ is a new acidic species. A new basic species Cl^- is also formed. These two new molecules are *ions* which carry a charge equal in magnitude to that of a freed electron. When a voltage is applied to this substance the ions carry a current between the poles of the conductor. In the case of the HCl solution the H_3O^+ ion is the *cation,* the CL^- ion is the *anion.* Cations are attracted to the negative poll of a conductor called the *cathode.* At the cathode electrons are yielded to the cation in a process called *reduction.* The anions are attracted to the positive pole, the *anode,* and are forced to yield electrons in a process called *oxidation.*

The double processes of reduction and oxidation which occur simultaneously at the cathode and the anode complete the electrical circuit and allow current to flow between the two poles. However, the process also involves chemical change. When the hydronium ion, the cation (H_3O^+), receives an electron at the cathode it loses its positive charge and changes into water (H_2O). Simultaneously when the anion (CL) yields an electron at the anode it reacts in a complex process with the water molecule to combine with hydrogen, reforming hydrochloric acid (HC1) and releasing free oxygen.

A similar process develops when a base is dissolved in water. In this case however the basic ion produced is called the *hydroxl* or *hydroxide ion* and water acts as an acid, attracting electrons.

The relationship between electrical current and chemical reaction in an electrolyte was specifically described by Faraday in 1833:

> The amount of chemical reaction taking place at an electrode is directly proportional to the amount of charge flowing through the electrolyte.

The process of electrolysis produces free hydrogen at the cathode and free oxygen at the anode in an acid electrolyte according to the following chemical process:

Cathode reaction:
4 electron charges + $4H^+$ (from hydronium ion) $\rightarrow 2H_2$

Anode reaction:

$2H_2O \rightarrow O_2 + 4H^+ +$ 4 electron charges

In a basic electrolyte the process operates accordingly:

Cathode reaction:

4 electron charges + $4H_2O \rightarrow 2H_2 + 4OH^-$

Anode reaction:

$$4OH^- \rightarrow O_2 + 2H_2O + 4 \text{ electron charges} \qquad \text{(Figure 4.2)}$$

The electrocatalytic process in an acid or base solution has been a fundamental means of producing hydrogen from electrical current for a number of years. Recently however new aspects of the process have been developed which promise to make the electrocatalytic production of hydrogen far more attractive. These processes have been developed in conjunction with the space program, particularly the research involving fuel cells. The development of these new processes was pioneered by The DuPont Company and the General Electric Company's work on a practical *Solid Polymer Electrolyte* (SPE).

General Electric's effort owes its considerable success to the development of a polymer called Nafion[R], a proprietary material of the DuPont company. A polymer is a molecule with a high molecular weight which is formed of smaller molecules. Polymers are formed from hydrocarbons composed of carbon and other elements, primarily hydrogen. Many familiar substances like rubber are polymers of

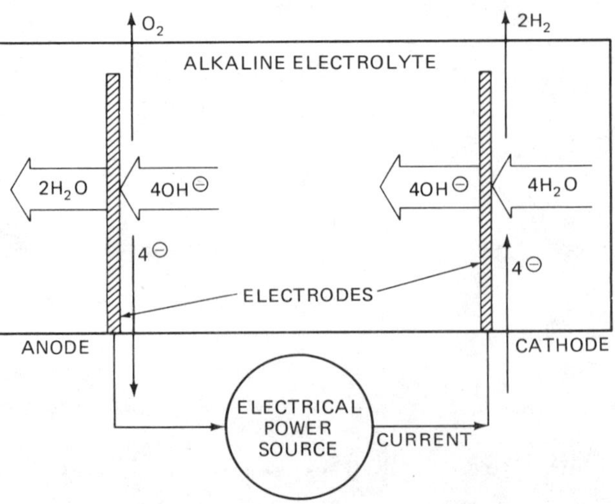

Figure 4.2 Hydrogen-oxygen electrolysis

other molecules. Rubber is composed of a large chain of molecules of a substance called *isoprene.*

Another polymer is the familiar DuPont substance *Teflon* which is a polymer of *tetrofluoroethene.*

Nafion^R is derived from Teflon. To make Nafion^R, teflon is restructured and a sulfonic acid group is attached. Nafion^R is a solid plastic substance which possesses unusual characteristics when made into thin sheets and immersed in pure water. The sulfonic acid groups (SO_3) remain stable in the structure. However, hydrolysis still takes place and the hydronium ion H_3O^+ is formed. This ion is mobile and passes through the sulfonic acid groups to carry a current through the thin material if a voltage is applied.

Solid Polymer Electrolyte can be fashioned into electrocatalytic devices to produce hydrogen. These devices possess a number of interesting advantages. The devices are formed essentially by coating each side of a thin sheet of Nafion^R with an inert catalyst material such as platinum which acts both as the anode and the cathode. This thin sheet is immersed in pure water and a voltage is applied between the anode and the cathode. Hydrogen is freed at the cathode and oxygen at the anode as in any acid electrolyte system. However the system has a number of additional characteristics. The SPE provides a physical separation between the resulting free hydrogen and oxygen so that there is a built-in barrier, essential for the collection of a gas. In addition, hydrogen can be produced at a high pressure for easier storage. The system is potentially simple, efficient at any current density, stable, and corrosion resistant.[10]

Regardless of whether an electrolysis system is of the membrane type just discussed or a water electrolysis system using a liquid electrolyte, certain considerations hold for all. A single unit of an electrolysis system is called a *cell.* Each cell must have at least four elements: an anode, a cathode, an electrolyte and a means of separating the hydrogen and oxygen called a *diaphragm.* In order to function each cell must have a source of electrical energy. Once this energy is applied to the cell it is manifested in two operational concepts called *cell voltage* and *current density.*

Cell voltage is a function of a number of concepts. The first of these is the theoretical minimum amount of electrical energy (E)

necessary for electrolysis to occur. This value results from a combination of the amount of energy necessary to decompose the water molecule, and the energy involved in the chemical processes at the anode and the cathode according to Faraday's law. A second concept is the amount of electrical energy being transmitted through the electrolyte expressed as *cell current* (I). The third concept is *cell resistance* (Ro). These three concepts are essential to any electrical circuit. Two other concepts are also involved; these are measures of the energy limitations of the anode and the cathode, Na and Nc respectively. These concepts can be operationalized in the formula:

$$\text{operating cell voltage } (V) = E + (Na + Nc + IRo)$$

Since high efficiency of a cell is achieved at low operating cell voltages, fundamental effort is devoted to reducing the Na, Nc and IRo values.

Current density on the other hand is a simple expression of cell current per unit area:

$$\text{Current density } (J) = \frac{I}{A}$$

Current density is useful in determining the potential monetary cost of a cell:

$$\textit{Module capital cost} = \frac{\text{(constant) (capital cost per unit area)}}{\text{current density } (J)}$$

The concepts of cell voltage and current density can be applied to any electrolysis system and provide valuable measures of cell efficiency and capital cost.*

Water electrolysis systems can be categorized according to whether the cells are electrically arranged in series or in parallel. Cells ar-

*The concept of cell voltage and current density, from which this description is taken, is explained in more detail in A. P. Ficket and Fritz R. Kalhammer, "Water Electrolysis," eds. Cox, Kenneth E. and K. D. Williamson, *Hydrogen: Its Technology and Implications,* Vol. 1, (Cleveland: CRC Press, 1977), pp. 6–9.

ranged in parallel are called *unipolar* systems. In these systems all cells share a common electrolyte. Systems which have cells arranged in series are called *bipolar* and each cell must have its own electrolyte. The solid polymer electrolyte modules can be used either in series or in parallel.

Unipolar systems have the advantage of being simple, reliable and flexible in that individual cells can be removed without shutting down the whole system. However, a large current is required for each cell and unipolar systems cost almost as much as bipolar systems, which are more expensive to produce but which have the advantage of occupying a smaller area for the same amount of hydrogen produced.[11]

Systems of both types produce hydrogen of great purity for use in chemical processes and are in use in a number of areas of the world. However, most of the hydrogen in commercial use is produced chemically using natural gas or other petroleum-related substances. The reason is a combination of the apparent abundance of petroleum and natural gas and the cost of electrical current. With the advent of cheaper electricity generated by solar energy and the development of new electrolysis technology, electrolytic hydrogen production is considered to be a major aspect of the model solar-hydrogen economy by the first quarter of the mature stage.

Two major areas of development in the production of electrolytic hydrogen are viewed as particularly important: *alkaline electrolyzer technology* and *solid polymer electrolyte technology*. The former technology is being pursued by Teledyne Energy System, Inc. and through improvements in gas separation is expected to result in greatly increased cell efficiency.[12] However, research and development efforts on large-scale hydrogen production in alkaline electrolysis systems have been underway for years. One system with the goal of developing a 2-megawatt electrolyzer was initiated in 1975 as a joint venture of Electricite de France (EDF) and Gaz de France (GDF). This system has been in operation on a research test basis since 1978 and a number of problems of large-scale operations have been overcome.[13]

The latter technology, which has been discussed, has the advantage of great simplicity and durability. The SPE system could be designed to function both as a fuel cell and as an electrolyzer (see Chapter 5). When combined with an effective photovoltaic device such a system could produce hydrogen from solar energy, thus storing energy, and later converting hydrogen to electricity for use when needed. These systems could be manufactured as modules which could be combined to make an energy system small enough for an individual home or large enough for an industry or a city.

Research and development efforts in the field of solid polymer electrolytic systems are rich and active because the potential of these systems is so exciting. The Brown Boveri Research Center in Switzerland, through its work on the MEMBREL cell, which uses Nafion^R as the electrolyte, is in the forefront of efforts to develop systems for the large-scale production of electrolytic hydrogen. In this system a platinum film deposited in the electrolyte surface is the cathode, and ruthanium and dridium covered by a current collector composed of porous titanium is the anode. This particular effort stresses the development of catalysts and current collections for improved efficiency.[14]

Another major research effort is the development of a high degree of chemical reversibility in solid polymer systems. The more reversible a system, the longer it will continue to function without replacement of elements consumed in the chemical processes. The oxygen electrode (the anode) is a particular source of irreversibility in solid polymer systems. General Electric's work on a new ruthenium-based, mixed-oxide catalyst, called WE-3, is a serious step toward increasing the operating life of solid polymer systems.[15]

It should be clear that research and development of solid polymer electrolysis systems is intimately connected with efforts to improve fuel cells. The two systems are in fact the reverse of each other. We will consider fuel cells in more detail in discussions of the conversion of hydrogen to energy in Chapter 5

We have discussed electrochemical systems which produce hydrogen through the interaction of the water molecule and electrical current supplied by solar energy external to the electrocatalytic cell. One of the more intriguing possibilities for the future however involves the combination of solar-generated electrical current and electrolysis in the same cell. The photoelectrolysis of water involves the suspension of photoelectrical materials such as very small particles of doped, single-crystal silicon in water solutions or other electrolytes. Problems of oxygen/hydrogen separation and long-term reliability have challenged the development of these systems for some time, but interest has remained high. We have previously discussed some of the more recent breakthroughs in the development of photovoltaic/hydrogen production systems by the Texas Instruments Corporation and researchers at Texas A & M University.*

Electrochemical conversion of energy to hydrogen combined with solar-produced electricity is one of the foundations of a mature solar-hydrogen economy. In the mature economy, hydrogen is the fundamental means of storing and transporting energy and as such it supplants all fossil fuels and long-distance transmission of electrical current.

We have considered three major ways of converting solar-produced energy into free hydrogen and within these three categories we have specified a number of technologies which are judged at this time to offer the best potential for implementing a solar-hydrogen economy. These judgments have been structured according to criteria which includes more than market competition. Among these criteria are long-range resource efficiency, impact on the environment, social and political effects, and technological simplicity.

The technologies which have been reviewed have been considered according to their position in the model. Coal gasification to hydrogen was selected as a primary hydrogen production process because it is a familiar technology, uses an abundant resource, and is controllable enough to prevent serious carbon dioxide and sulfur dioxide pollution.

The production of hydrogen by electrolysis is an elegant technology with application in all stages of the model. However, except for combined purposes, such as the generation of hydrogen and electrolytic ocean mining, systems which use liquid electrolytes are considered applicable in only the first stage of a solar-hydrogen economy. The middle period and the mature economy depend increasingly on solid-polymer electrolysis systems.

Finally, the two technologies which appear to have great long-range potential but which require considerably more development are those which involve the thermal decomposition of water and those which involve photosynthesis and photosynthesis mimicking. These two technologies are elegant and potentially efficient, but they both involve serious technical difficulties and safety concerns which, while

---

*See Chapter 2.

deemed solvable, place them beyond the possibility of rapid implementation.*

NOTES

1. Peter Hoffmann, *The Forever Fuel, The Story of Hydrogen.* (Boulder: Westview Press, 1981).
2. L. D. Hadder. "The Economics of Producing Hydrogen from a Small Air-Blown Coal Gasifier" (The Billings Energy Corporation, unpublished manuscript).
3. For an excellent review of the current status of photobiotics, see Paul F. Weaver, Stephen Lien, and Michael Seibert, "Photobiological Production of Hyrodgen," *Solar Energy,* Vol. 24, November 1, 1980, pp. 3–45.
4. Ibid.
5. A. I. Krasna. *Biological and Synthetic Systems for Production of Hydrogen from Water.* (New York: Columbia University for National Science Foundation, Washington, D.C., March 1980).
6. Weaver, p. 24–26.
7. For a brief description of some recent research see "Hydrogen Fuel," *The Economist,* March 27, 1982, pp. 92–93.
8. Hoffmann, p. 69.
9. For a brief description of solar hydrogen production high temperature systems see E. Bilgen, "Solar Hydrogen Production at High Temperature," in *Solar Energy Conversion,* ed. A. E. Dixon and J. D. Leslie (Toronto, Canada: Pergamon Press, 1979), pp. 1137–1144.
10. W. J. D. Escher et al., *Solar/Hydrogen Systems Assessment,* U.S. Department of Energy, DOE/JPL-955492, Vol. 2, Part 2, June 2, 1980, pp. 37–40.
11. Cox.
12. E. Fein, "The Market Potential for Electrolyic Hydrogen," *International Journal of Hydrogen Energy,* Vol. 6, No. 1, Pergamon Press Ltd., 1981, pp. 57.
13. C. Bailleux, "Advanced Water Alkaline Electrolysis: A Two Year Running of a Test Plant," *International Journal of Hydrogen Energy,* Vol. 6, No. 5, Pergamon Press Ltd., 1981, pp. 461–471.
14. S. Stucki and R. Muller, "Evaluation of Materials for a Water Electrolyzer of the Membrane Types," in *Hydrogen Energy Progress, Proceedings of the 3rd World Hydrogen Energy Conference* T. W. Veziroglu et al., (Tokyo: June 23–26, 1980), pp. 1799–1808.
15. J. M. Sedlak et al., "Advances in Oxygen Evolution Catalysis in Solid Polymer

*For a basic review of some of the principles and potential of the solar photoelectrolysis of water see A. F. Janzen, "Photoelectrochemistry I—Photoelectrolysis," *Solar Energy Conversion,* (Toronto: Pergamon Press, 1979) A. E. Dixon and J. D. Leslie, eds. and R. F. Schwerzel et al., *Evaluation of the Commercial Feasibility of Hydrogen Production by the Solar Photoelectrolysis of Water Task One; Interim Report,* Columbus, Ohio: Battelle Columbia Laboratory for U.S. Department of Energy, May 1977.

Electrolyte Water Electrolysis," *International Journal of Hydrogen Energy,* Vol. 6, No. 2, Pergamon Press Ltd., 1981, pp. 159–165.

REFERENCES

Andersen, Larry L. and David A. Tillman. *Synthetic Fuels from Coal.* New York: John Wiley and Sons, 1979.

Bailleux, C. "Advanced Water Alkaline Electrolysis: A Two Year Running of a Test Plant." *International Journal of Hydrogen Energy,* Vol. 6. No. 5. Pergamon Press Ltd., 1981.

Bilgen, E. "Solar Hydrogen Production at High Temperature," *Solar Energy Conversion.* eds. A. E. Dixon, and J. D. Leslie. Toronto: Pergamon Press, 1979.

Cooper, Bernard R. *Scientific Problems of Coal Utilization.* Conference Proceedings. U.S. Department of Energy, 1978.

Cox, K. E. and K. D. Williamson, Jr. eds., *Hydrogen: Its Technology and Implications.* Vol. 1. Production Technology. Cleveland: CRC Press, Inc., 1977.

Escher, W. J. D. et al. *Solar/Hydrogen Systems Assessment.* U.S. Department of Energy. DOE/JPL-955492. Vol. 2. Part 2. June 2, 1980.

Fein, E. "The Market Potential for Electrolyic Hydrogen." *International Journal of Hydrogen Energy.* Vol. 6. No. 1. Pergamon Press Ltd., 1981.

Hadder, L. D. *The Economics of Producing Hydrogen from a Small Air-Blown Coal Gasifier.* The Billings Energy Corporation. Unpublished manuscript.

Janzen, A. F. "Photoelectrochemistry I—Photoelectrolysis" *Solar Energy Conversion* eds. A. E. Dixon and J. D. Leslie. Toronto: Pergamon Press, 1979.

Krasna, A. I. *Biological and Synthetic Systems for Production of Hydrogen From Water.* New York: Columbia University for National Science Foundation, Washington, D.C., March 1980.

Pelofsky, Arnold H. ed. *Coal Conversion Technology, Problems and Solutions.* Washington, D.C.: American Chemical Society, 1979.

Schwerzel, R. F. et. al. *Evaluation of the Commercial Feasibility of Hydrogen Production by the Solar Photoelectrolysis of Water Task One: Interim Report.* Columbus: Battelle Coumbia Laboratories for U.S. Department of Energy, May 1977.

Sedlak, J. M. et al. "Advances in Oxygen Evolution Catalysis in Solid Polymer Electrolyte Water Electrolysis." *International Journal of Hydrogen Energy.* Vol. 6. No. 2. Pergamon Press Ltd., 1981.

Stucki, S. and R. Muller. "Evaluation of Materials for a Water Electrolyzer of the Membrane Types." in *Hydrogen Energy Progress, Proceedings of the 3rd World Hydrogen Energy Conference* Tokyo: June 23–26, 1980.

The Economist. "Hydrogen Fuel", March 27, 1982.

Veziroglu, T. Nejat ed. *Alternative Energy Sources II, Volume 7, Hydrocarbon Conversions.* Washington, D.C.: Hemisphere Publishing Corporation, 1981.

Weaver, Paul F. et al. "Photobiological Production of Hydrogen." *Solar Energy.* Vol. 24. Nov. 1, 1980.

5. HYDROGEN TO ENERGY

Electrochemical conversion of hydrogen to electricity in fuel cells and combustion of hydrogen are two processes fundamental to the solar-hydrogen energy economy. Of these two conversion processes electrochemical conversion is considered to be a foundation element of a mature solar-hydrogen economy. Combustion of hydrogen on the other hand is an important intermediate process.

ELECTROCHEMICAL CONVERSION—THE FUEL CELL

We have discussed the process of electrolysis in which the water molecule is dissociated into hydrogen and oxygen when an electrical current is passed through the free ions of an aqueous electrolyte. This process can be reversed. If hydrogen is introduced at the electrolyte anode it will combine with hydroxyl ions to form water and release free electrons according to the following formula:

$$H_2 + 2OH^- \rightarrow 2H_2O + 2 \text{ free electron charges.}$$

If oxygen is also introduced at the cathode along with an electron charge it will react with water to form hydroxyl ions as follows:

$$2H_2O + O_2 + 4 \text{ free electron charges} \rightarrow 4OH^- \qquad \text{(Figure 5.1)}$$

If these reactions take place independently of each other, even in the same electrolyte, they act as totally separate chemical processes and rapidly proceed to equilibrium. If however the anode and the cathode are connected by a conductor of electricity then the free electrons will transfer their energy from the anode to the cathode through the conductor thus sustaining the formation of hydroxyl

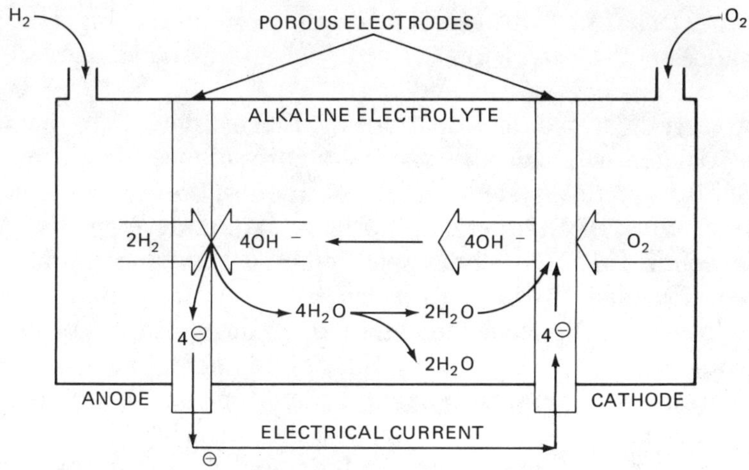

Figure 5.1 Hydrogen-oxygen fuel cell

ions at the cathode. These free ions then move through the electrolyte to complete the circuit. The products of the complete process are simply electrical current, pure water and heat. The device for developing the above process is called a fuel cell.

The use of fuel cells to generate power for domestic consumption has attracted interest since the turn of the century, but the effort to initiate widespread use has not been undertaken until fairly recently. The lack of fuel cell application is due in part to the development of apparently inexpensive combustibles like coal, natural gas and petroleum.

The potential for fuel cell use seems to have been lost in the wake of the rapid industrial expansion which relied on the combustion of hydrocarbons. By 1956 research in fuel cells was being conducted by only a few persons throughout the world. Bacon in England, Justi in Germany, Davtiyan in the USSR, Ketelaai in Holland and Gorin in the U.S. were pioneers who kept the effort alive.[4] However, in the five years between 1956 and 1961 research into fuel cells increased dramatically, and by 1961 corporate activity in the field had increased from 5 to over 50 in the U.S. alone. These firms represented industries from public utilities to aircraft.[5]

With the exception of application for the space program, most research into fuel cell development has been directed toward the use of fuels other than free hydrogen. The primary reason for this is the predominance of the petroleum-derived fuel and associated fossil fuel systems such as natural gas and coal.

Very early efforts to develop domestic fuel cells highlight the problems associated with this approach. An early pioneer in the attempt to develop a practical system was W. A. Jaques. Basing his efforts on the earlier work of Becquerel and others, Jaques developed a working fuel cell in 1897. His model operated on carbon and air. The fuel however was really hydrogen formed by an electrolyte-anode reaction.[6] This early experience indicates a dilemma which has been present throughout the history of fuel cell development. Jaques' attempt to use carbon as a fuel was to be duplicated time and again up to the present. The reason for this duplication is clear. Throughout human history carbon compounds, particularly hydrocarbons, have been thought to be the only practical fuels available. The only possibility for the use of fuel cells therefore seemed to lie in their capability to utilize hydrocarbons as fuel. The potential of fuel cells could only be realized if techniques were developed to separate hydrogen, which is the actual fuel, from the hydrocarbon compounds. Because of this need for separation of hydrogen, the history of the research and development of fuel cells for terrestrial use is really the history of two separate technologies put together in one system. The first of these technologies is the production of hydrogen from hydrocarbons; the second is the conversion of the energy in the free hydrogen molecule into electrical current.

While the principles of fuel cell operation are simple, the development of a practical model has been a long process in which a number of engineering problems have been confronted. These problems can be roughly categorized into those involving the introduction to hydrogen and oxygen into the electrolyte, those involving the removal of heat and water, and finally the structural design of the cell to conserve size and weight.

Hydrogen and oxygen can be supplied to the cathode in a number of ways and in several states of chemical freedom. While the reactants in this process are most frequently hydrogen and oxygen, these reactants need not be chemically free when they are introduced into the cell. Air can be used in place of pure oxygen in which case oxygen is chemically extracted in the reaction at the cathode and the remaining gases are by-products which must be considered as wastes.

Hydrogen in its chemically free state is the most efficient fuel. However, much fuel cell research has been devoted to developing technologies for using hydrocarbons such as methane CH_4, ethylene C_2H_4, and propane C_3H_8. These substances are fundamental aspects of the petroleum industry, they are readily available at present, and the carbon-combustion economy is organized and structured to make their transportation and storage more convenient than free hydrogen. The fundamental problem with using hydrocarbons as well as other compounds as fuels is that hydrogen must be chemically released from them before the reaction at the anode can take place. This requirement means that cells which use hydrocarbons as fuel are really two devices. They are devices for releasing hydrogen from chemical bonding, and they are devices for converting free hydrogen to water and energy. When these two requirements are combined in one device its efficiency as an energy producer will obviously be decreased by the amount of energy required to release hydrogen from its chemical bonding in hydrocarbon or alcohol. Regardless of what fuel is used it must be introduced at the electrode in a way which will encourage the necessary chemical reaction to take place.

Introduction of oxygen and hydrogen into the fuel cell through porous electrodes was an early approach to the problem of uniting the reactants with the electrolyte. This however was not an efficient way of handling the problem because of the difficulty involved in assuring an even distribution of gas along the surface of the electrodes. A basic method for overcoming this difficulty was to place the gas in a reservoir behind a porous electrode formed in a thin plate. This method allowed the gas to be evenly distributed over the entire area. The technique was also compatible with a design feature which helped solve the problem of ion travel distance through the electrolyte. The less the travel distance the better, so it made sense to design the cell as a thin wafer with the electrolyte sandwiched between the two plate electrodes which were themselves faced by the gas chambers.

The second problem involves the removal of by-products of fuel cell operation. Electricity, water and heat are generated in fuel cells.

All three of these products have potential uses. However, water and heat are often considered to be wastes. Heat can be removed either by natural radiation or by the circulation of a coolant. In small units radiation is an adequate method of heat removal but as the number of modules increases radiation does not suffice and coolants must be circulated. The coolant can be the electrolyte itself, the fuel, or a separate substance circulated through the system. All three of these processes have been used in various devices. Water is a problem because as it is produced it becomes part of the electrolyte and can, over time, dilute the electrolyte beyond usefulness. There are a number of ways of removing the water produced during fuel cell operation; many of the techniques now employed depend on the pressure variation in cells.

The problem of structural design, as with most fuel cell research, has been closely tied to the space program. Space vehicles needed a source of power which was independent, flexible, had high operational efficiency, and consumed a relatively small amount of fuel. After a review of all possible types of power units the National Aeronautics and Space Administration (NASA) decided on fuel cells. The fact that they produced water as a by-product was an important added benefit.

The fuel cell used in the *Apollo* space craft was based on a system developed by Francis T. Bacon in 1932. This cell used hydrogen and oxygen, an alkaline electrolyte, and metallic nickel electrodes. Bacon's original system operated at high pressure (600 psi) which provided a considerable improvement in efficiency over models operating at atmospheric pressure. The construction requirements to allow for such a high pressure however made the system too heavy for spacecraft so the Apollo cell (developed and constructed by Pratt and Whitney Aircraft) was designed to operate at lower pressure (50 psi). In order to keep the electrolyte from boiling at this pressure the concentration of alkaline (KOH) in the electrolyte was increased to 75%. The temperature was raised to 260°C in order to increase the efficiency of the cell. The cell module weighed about 245 lbs., and operated over a normal power range of 523 to 1420 watts at a normal voltage of about 30 volts. It was 44 inches high and 22.5 inches wide.

Three of these modules connected in parallel provided electrical power and water for the Apollo space craft. Each module was composed of 31 cells connected in series with associated control systems.[1]

The system used in the *Gemini* space craft was quite different and was based on a more advanced technology. Instead of a water-alkaline electrolyte operating at high temperature, the Gemini system used a solid electrolyte made of sulphonated polystyrene. This substance had been pioneered by the DuPont Company and prepared as a proprietary product called Nafion[R] (see Chapter 4). A lightweight, organic structure, this solid polymer electrolyte had many advantages over the liquid. Simple handling, stability and well-defined boundaries which did away with problems of electrolyte storage made it ideal for use in fuel cells.

The system was composed of two sections, each of which contained three stacks of 32 cells. Each section was 25 inches long and about 12 inches wide so that the whole system measured about two cubic feet. Each section produced 1000 watts at about 24 volts and each stack could be removed as desired, which gave the system great flexibility.[2]

If it is difficult to speculate on the state of the art of fuel cells in the space program, it is almost impossible to do so with regard to military application. It is however safe to say that considerable research has been conducted and that a number of new systems have been developed. Some of these systems, we know, follow the research of earlier models. Both General Electric and Pratt and Whitney appear to be active. In 1973 G.E. was developing a new system which weighed between 10 and 30 lbs. This system was designed to operate from 5,000 to 10,000 hours and produce 6 to 10 kW. A Pratt and Whitney system was heavier (85 lbs.), but it was claimed to produce up to 50 kW for short periods.[3]

Research and development associated with the space program and the defense industry can be expected to continue to be the leading edge of fuel cell technology development, particularly with regard to decreasing the size/output ratio. However, in recent years there has been a great increase in efforts to demonstrate the potential of fuel cell application to terrestrial energy needs.

Applications: Electrochemical Conversion Devices

The possible applications of fuel cell technology on earth are many and varied. Fuel cells can power automobiles and rail transportation systems. They can provide electricity, heat and cooling for homes, and they have the long-range potential for supplying energy for industry and the simultaneous production of fresh water. When combined with the solar production of hydrogen, fuel cells are a potential source of energy, which is both nonpolluting and totally renewable. In this application however the term fuel cell may be somewhat misleading for it must be remembered that a fuel cell can also be an electrolyzer if the process is reversed. When a fuel cell is operated in this manner it is often referred to as a *regenerative fuel cell.* A better term for a device that operates both as a fuel cell and an electrolyzer might be *electrochemical conversion device* (ECD) and we will use this term for the remainder of our discussion. An electrochemical conversion device when operated in conjunction with a photovoltaic device and a hydrogen storage system could provide the total energy needs for residential and commercial units and in addition produce fresh water.

Perhaps the best example of this research and development direction, though by no means the only one, is a program begun in 1967 by a consortium of gas utility companies. This program called TARGET (Teams to Advance Research for Gas Energy Transformation) set as its goal to develop, over a nine-year, three-phased program, a fuel cell energy service. The prime contractor for this effort was the Pratt and Whitney Company which, as we have noted, had considerable experience in fuel cell research.

The program resulted in the design of a system to convert methane into hydrogen and carbon dioxide, using the hydrogen in a fuel cell, and finally to convert the DC current to AC in an inverter. The system was designed to provide other energy benefits such as heating, cooling, and humidification.[7] A plan for the development of fuel processors, or converters, was developed for the U. S. Department of Energy in 1978.[8] This plan was based on the use of phosphoric acid and molten carbonate electrolyte fuel cells which use impure hydrogen and air, thus the conversion process is not designed to produce pure hydrogen. The converter in this case involves a two-stage process of water reforming and shift conversion. Water reforming is primarily the reaction of a hydrocarbon with water. The reaction for methane is similar to that for methanol, naphtha or basically any other hydrocarbon:

$$CH_4 + H_2O \rightarrow CO + 3H_2$$

The shift conversion proceeds according to the following formula:

$$CO + H_2O \rightarrow CO_2 + H_2$$

It should be noted that these reactions are basically similar to the reactions in the process of caol gasification to hydrogen.

The system which this plan uses as a point of departure is a conversion process developed by United Technologies Corporation (UTC) for demonstration units of 4.8 MW and 40 kW. The former unit would be the size useful in a public utility; the latter would be applicable for commercial or multiresidential uses. While the plan does not address single residential applications, it does confront directly the issue of the separation of the technologies of hydrocarbon conversion to hydrogen and the use of hydrogen in fuel cells.

The plan for developing fuel processors is directed toward the use of fuel cells to produce electricity for electrical utilities; however, it clearly recognizes that this approach to the production of electricity is in opposition to the prevailing philosophy of utility power generation. As the plan notes, utilities have generally adopted a strategy of constructing "large-scale base-load plants" which use smaller, less efficient power units to produce extra electricity during hours of peak demand. This strategy is highly centralizing, but the use of hydrocarbon converters and fuel cells could lead to the decentralization of power production. The plan recognizes this dissonance and notes that there may be institutional barriers to the development of such fuel cell systems. It further notes that these institutional barriers will have to be overcome before smaller fuel cell systems for domestic residential, commercial or industrial purposes can be successfully deployed in the market.[9]

Another Department of Energy study conducted by the Oak Ridge National Laboratory projects that there can be an installed capacity of 3.4 million kW of energy produced by fuel cells in the United States between 1985 and 2000. This projection is based on a series of assumptions which include the exercise of no institutional, legal, or regulatory barriers. In addition, the projection assumes that at least four sizes of fuel cell modules will be available (based on the phosphoric acid cell and the development of the UTC converter) in the following ranges: 25, 50, 100, and 250 kW. The study projects that the system will be utility owned, and that the only public inducement would be a 10% business investment tax credit.[10]

The assumptions which led to these projections have not been realized since the study's publication due to a major shift in the federal government's policy toward energy. The potential for the development of the projected market penetration remains, and it could be realized with some adjustments in the projected time frame.

The introduction of fuel cell technology for residential, commericial and industrial use through the electrical utility system is a major and important step in the early stages of implementing a solar-hydrogen economy. Such a step could result not only in the further development of the technology but also in a clarification of the separate natures of technology for converting hydrocarbons to hydrogen and the technology for converting hydrogen to electricity. However, it must be recognized that in such systems the natural gas and methane are not really fuels, but only hydrogen storage systems. One question which surely must be confronted is whether the hydrogen should be separated from the hydrocarbon at the point of use or at some central location and then piped to the consumer for use in an onsite fuel cell. The answer to this question depends on a number of considerations.

The present carbon-combustion energy economy certainly favors point-of-use separation such as the TARGET system. Such systems take advantage of the existing natural gas network of distribution and do not seriously threaten the present energy economy. These systems however tend to distort the potential of fuel cell technology. Because they combine separation technology with electrochemical conversion technology, these two processes are lumped together when efficiency and cost are calculated. As a result the consumer never realizes the greater efficiency of electrochemical conversion over combustion. In addition, these systems are inexorably tied to the hydrocarbon production of hydrogen. As a result of their unique design they are unable to utilize hydrogen produced by other means.

Another problem with this approach to fuel cell development is that it tends to mask the tremendous potential flexibility of electrochemical conversion for domestic use. In 1975 when J. O'M. Bockris first outlined the potential of a hydrogen economy he had some of this flexibility in mind and he described a number of possibilities.[11] Since that time the potential has increased. Much of this increased potential for electrochemical conversion is due to developments in photovoltaics and other forms of alternative solar power. The long-range potential benefits to be realized through the domestic residential use of electrochemical conversion devices largely depends on the development of systems which can take full advantage of the various technologies for producing hydrogen.

Systems which link electrochemical conversion devices (ECD) with hydrogen production systems could be used in every segment of the energy economy. In the domestic residential segment the system could be decentralized to the extent that individual homes could produce, store and use energy independent of an outside source. Individual photovoltaic devices could convert sunlight to direct electrical current and the current could be used to produce hydrogen in an ECD. The hydrogen could be stored in metal hydrides and later reconverted into electricity in the same ECD operating as a fuel cell. The system would be independent, silent, use no fuel but sunlight and water, and produce virtually no pollution.

Mixed centralized systems could be established which would provide electrical energy for communities and industries. These systems could use electricity produced by any solar energy system. Such systems could take advantage of the unique nature of the particular area by using wind, wave, tidal OTEC or photovoltaic electrical generation. Electrochemical conversion device modules could be "stacked" to provide any size conversion process. In such systems hydrogen could be stored in large tanks, metal hydrides or other ways depending on the needs and capabilities of the demand area.

These systems could produce hydrogen and oxygen by electrolysis during periods when electrical current is available (when the demand for utility power is low or during the operation of photovoltaic devices). Hydrogen could be produced and stored while the system operates as an electrolyzer. During periods when electrical current is not available, at night or during peak utility hours, such systems could operate as fuel cells and use the stored hydrogen to generate electrical current. Work on such cells as these has been sponsored by NASA for a number of years. Among the companies which have been active in this effort are Pratt and Whitney and Energy Research Corporation.

Applications: Transportation

Automobiles have been built which use fuel cells to generate electricity and then use this electricity to run the motor. An early and very ambitious automobile project was begun in 1964 by the General Motor Corporation. GMC constructed a fuel-cell powered van which used the most advanced technology available at the time. The fuel cell system used in the van was developed by Union Carbide. It weighed 3380 lbs. and developed 160 kW peak power with a rated output of 32 kW. The van used an advanced electrical motor which operated at 13,700 rpm. While the system was never road-tested its laboratory performance showed a top speed equivalent to a standard van but a range and acceleration only 60% that of a gasoline powered auto. The thermal efficiency of the system was about twice that of a gasoline engine. Problems with the system included: heavy weight and large size (about twice the weight of a regular van), short lifetime, high cost, complicated operation and complexity.[11] Bernard Crowe notes that many of these problems could probably have been overcome if the project had not attempted to duplicate the performance of a standard large gasoline powered vehicle.

A much more successful project was that of Dr. Karl Kordesch. Kordesch built an electric "city car" which used a fuel cell supplemented with a battery. His car weighed only 2000 lbs. and had a range of 200 miles. Most important, this system, while small in comparison with GMC's van, has successfully operated on public roads for some time.[12]

Neither of these systems used the advanced technology of fuel cell design which is available today. Fuel cells which use solid polymer electrolytes and hydrogen stored in metal hydrides could add a degree of reliability and safety unknown in these early designs.

It is generally acknowledged that fuel-cell powered autos are a very important long-term possibility. It is also generally accepted that near-term transition to this stage will involve the combustion of hydrogen or methanol in internal combustion engines.[13] This transition stage will involve the further refinement of fuel cells and the development of lighter, safer hydrogen storage systems.

Another important, more immediate use of fuel cells is their application to rail transportation systems. Modern diesel engines actually use electric motors to provide their motive power. The diesel engine is used to turn a generator which makes the electricity used by the motor. Fuel cells could be used to generate this electricity much more efficiently. Since weight and start-up time are not really problems in rail transportation, present generation fuel cells could be used. In addition large amounts of metal-hydride-stored hydrogen could be stored in tank cars with little concern for their weight. The safety problems associated with such storage would be less than those associated with the storage of diesel fuel.

ECDs operating as fuel cells in transportation systems have very great potential. This potential is a function of their relatively greater efficiency over internal combustion engines when they are combined with electric motors to make a complete engine. Fuel cells appear to have real application in both automobiles and in railway systems. The potential inherent in such flexibility would seem to argue for the development and marketing of a reliable and efficient ECD which utilizes hydrogen and air and could operate both as an electrolyzer and as a fuel cell.

Systems of this type, while simple and elegant in concept, pose great technological challenges. A great problem with these devices is that the electrolysis phase of operation and the fuel cell phase cannot both be individually maximized in the same device.[14] The basic research challenge would thus seem to be the need to balance the device so that the maximum possible efficiency is achieved in each phase while whole system stability is preserved. The development of solid polymer electrolytes and the research that these and other breakthroughs are stimulating should greatly encourage the development of reversible electrochemical conversion devices.

Advantages of Electrochemical Conversion

We have examined some of the possibilities and problems associated with the use of ECDs for residential, commercial and industrial purposes. There are many advantages to the application of electrochemical conversion which should be emphasized. Some of these advantages are apparent when a simple comparison is made with the method of producing electricity through the combustion of hydrocarbons such as coal, oil or natural gas. The production of electricity in a modern day utility is essentially a three-stage process. First the hydrocarbon is burned to make heat, then the heat is used to boil water, and the steam from the water is used to turn a turbine which runs an electrical generator. There is a loss of energy during each stage of this process. Electrochemical conversion devices use a one-step process to generate electrical current. This process is primarily a result of the migration of ions through an electrolyte. There is thus no waste in the transition from heat energy to mechanical energy to electrical energy.

In addition to the evident advantages inherent in the comparative simplicity of the system there is an efficiency advantage in the electrochemical process over the combustion process. The maximum theoretical efficiency of a hydrogen-oxygen electrochemical conversion device in the electricity-producing mode (a fuel cell) is 83%. The Carnot cycle, which demonstrates the efficiency of combustion, indicates that no such an energy conversion efficiency is possible for the combustion of hydrogen and oxygen even at temperatures as high as 2000°K. In practice the theoretical efficiency of fuel cells is not reached due to the high irreversibility of the chemical reaction at the oxygen electrode. Research efforts into ways of more closely approximating the theoretical efficiency are being conducted.[15] The potential efficiency of electrochemical energy processes is clearly much greater than combustion.[16]

Another great advantage of the electrochemical conversion device is that it can be used to disperse the production of electrical energy thereby reducing the distance electricity must be transported. This advantage becomes very important when it is realized that much electricity can be lost in transmitting.[17] Amory Lovins estimates that as much as 40% of generated electrical current may be lost each year in transmission.[18] Hydrogen and other gases can be transported with almost no loss. ECDs can take advantage of the efficiency of transporting energy stored in gases. They can be built in small modules which can be used in individual homes or grouped together and used in communities or industries.

Finally, and certainly not least, electrochemical conversion devices do not produce wastes which are environmentally dangerous. The hydrogen-oxygen device makes water, electricity and heat. When air is used in place of oxygen the wastes are largely benign.

These advantages combined with the possible long-range integration of electrochemical conversion devices with onsite generation of electricity using photovoltaic systems make ECDs a central technology in a mature solar-hydrogen economy. In order to realize these advantages however considerable research and development is needed.

As previously mentioned one major ongoing research and development effort is directed toward the development of the phosphoric acid fuel cell for use by utilities. Such a cell must be able to operate for at least 10,000 hours reliably and produce electricity at a rate of $400/kW according to a recent study by the Arthur D. Little Company.[19] The next step in the process of developing a fuel cell designed for use by public utilities might be the completion of a cell which uses molten carbonate.[20]

This particular research and development effort, as previously mentioned, is directed toward the design of a fuel cell device which uses hydrogen in an impure form. In order to use naphtha, methane or other fossil fuels these cells must first "crack" the hydrocarbon into hydrogen and carbon monoxide. These two gases then become the fuel. Fuel cells such as these operate of necessity at high temperatures, up to 1200°K, which precludes the use of aqueous electrolytes.

Efforts to develop a fuel cell technology which uses hydrocarbons are tied to the carbon-combustion economy, and, from the perspective of the development of a solar-hydrogen economy, must be viewed as having at best transitional application.

Systems which use pure hydrogen and which are not required to convert hydrocarbons to hydrogen can be twice as cost effective as those which use hydrocarbon fuel. These systems can also be linked with any of the hydrogen-producing technologies.* However, serious research and development problems must be solved before such systems will be economically competitive with combustion. It will be extremely difficult to confront these problems in the context of the carbon-combustion economy because of the lack of fundamental economic incentives. Nevertheless these problems must be overcome if a solar-hydrogen economy is to be realized.

There are two major research efforts which must be encouraged in order to develop hydrogen-and-air electrochemical conversion devices. The first and perhaps most fundamental of these involves the high cost of the catalysts (often noble metals) which must be used in the electrodes. This is a particularly serious problem in systems which use air at one of the electrodes. There are at least three research directions which have been undertaken to solve this problem: the use of cheaper catalysts, more efficient use of catalyst material, and the use of less catalyst. Work is progressing in each of these areas.[21]

The second major problem area is that of system design. The challenge in designing a hydrogen/air fuel cell is primarily one of balancing the requirement of high chemical reactivity with that of long life and reliability. An optimum balance between these two frequently conflicting demands will be necessary in the development of an efficient, reliable and cost-effective electrochemical conversion device.[22] Effective research and development of hydrogen electrochemical conversion devices for terrestrial uses is inherently tied to research and development of the other basic technologies of the solar-hydrogen economy. The potential benefits of these research and development efforts cannot be recognized unless they are viewed in a whole system context together with other basic technologies for the conversion of energy to hydrogen and the transportation and storage of hydrogen.

---

*For an interesting discussion of fuel cell application in a solar-hydrogen context using OTEC see James R. Roney, "OTEC to Hydrogen Fuel Cells—A Solar Energy Breakthrough," *Hydrogen Energy Progress, the Proceedings of the 3rd World Hydrogen Energy Conference,* Vol. 3 (Tokyo, Japan, June 1980).

COMBUSTION CONVERSION OF HYDROGEN TO ENERGY

The production of hydrogen by solar energy and conversion of hydrogen to electricity in electrochemical conversion devices are the heart of a mature solar-hydrogen economy. The development of this system as a primary energy pattern is viewed as a process which could require almost a century to complete. Among the reasons for this rather long phasing process is the need to establish clearly defined, dependable and safe hydrogen storage and distribution patterns. We have also seen that while the basic technology for Electrochemical Conversion Devices is already available there is considerable research and development effort yet required to make these devices as reliable and as cost effective as their principle of operations promises they can be.

These two conditions point toward a strategy for implementing a solar-hydrogen economy which includes the introduction of hydrogen into general use in ways most compatible with present-day energy systems. Two aspects of the carbon-combustion economy seem to offer the best opportunity for the introduction of hydrogen. The first of these is the domestic residence. The energy for heating, cooling, and cooking could be supplied through the combustion of hydrogen relatively soon in some areas, and it would be possible in the middle range to supplant almost entirely the use of natural gas with hydrogen. The second aspect of the carbon-combustion economy which could readily accommodate the use of hydrogen is the transportation sector. The combustion of hydrogen rather than hydrocarbon fuels has many advantages in all three major aspects of the terrestrial transportation system. Hydrogen has been used very efficiently in jet aircraft engines as well as in internal combustion engines, and it promises to provide efficiencies for use in railway locomotives.

Moreover, the storage and transportation systems for hydrogen are essentially the same whether it is used through combustion or electrochemical conversion. These systems can be developed in the short-term while hydrogen is introduced for combustion uses compatible with present-day technology. At the same time research and development of electrochemical conversion devices as well as solar energy systems can be pursued and incrementally introduced as they become available.

How Hydrogen Burns

Hydrogen burns differently then hydrocarbons. The differences represent efficiencies, in general, which favor the use of hydrogen over hydrocarbons for residential and transportation purposes. However, they also require different safety and handling procedures.

All chemical reactions involve energy negotiations as we noted earlier. The energy negotiated in chemical reactions is the difference between the energy necessary to maintain the previous molecular state and that required to maintain the new one. When hydrogen combines with oxygen it is *oxidized,* and the energy necessary to maintain individual molecules of hydrogen (H_2) and oxygen (O_2) is greater than that necessary to maintain the molecule of water (H_2O) which they combine to make. This energy is released on combination. In order for the process to take place a relatively small amount of external energy must be applied. We have noted that when this external energy is electrical then the energy released in the reaction is also electrical. It is this reaction which drives an electrochemical device in the fuel cell mode of operation.

If the external energy is applied in the form of heat then the reaction will take place more rapidly and the energy released will be in the form of heat rather than electrical current. This combustion process for hydrogen and oxygen can result in temperatures as high as 2800°C. The heat from this process can be used to run internal combustion engines, provide warmth for homes and businesses, and supply the full range of uses to which fire is usually put. The process has one great advantage; when hydrogen is burned in air it produces only heat, fresh water and controllable amounts of nitrogen oxides. Since hydrogen is a renewable resource readily available from a number of sources the combustion of hydrogen would seem to be preferable to the combustion of hydrocarbons such as natural gas (primarily methane, CH_4) and gasoline which are limited resources that produce serious pollutants when they burn.

All substances that burn have different chemical characteristics which makes the products of their combustion different. Since our present carbon-combustion economy is organized and designed to burn hydrocarbons certain relatively minor adjustments have to be made in the system in order for it to accommodate hydrogen. These adjustments are no more difficult to make than those which were required to make the transition from the use of coal and wood to the use of natural gas. A comparison of the relative characteristics of hydrogen and hydrocarbons is necessary for an understanding of these accomodations.

Hydrogen is the lightest of all elements. In its free form it exists as a colorless gas. When equal weights of hydrogen and gasoline are compared with regard to their relative heating values, hydrogen is almost three times as high. However under normal circumstances a particular volume of hydrogen weighs less than a equal volume of either natural gas or gasoline. The heating values of a particular volume of natural gas will therefore be about 3.26 times as great as an equal volume of hydrogen.

Another important difference results from the fact that the combustion of hydrocarbons requires considerably more air than hydrogen. The fuel/air ratio (stoichiometric ratio) of natural gas (primarily methane) is 1:10 whereas the ratio necessary to receive the highest energy output for hydrogen is about 1:2.4.[23]

Finally hydrogen differs from hydrocarbons with regard to the rate at which it burns and the amount of heat necessary to ignite it. Hydrogen burns about 16 times faster than methane when both are ignited at their best fuel/air ratios. Hydrogen has a low ignition energy in comparison with methane, but it has a slightly higher ignition temperature. The temperature of ignition of hydrogen is such that a lighted cigarette would not ignite a hydrogen and air mixture. However, a lighted match would.

A final important difference between hydrogen and methane is that hydrogen mixes with air much more rapidly. This greater diffusion rate is important because it indicates another reason why hydrogen will reach a combustible mixture with air more rapidly than will methane. Thus the combustion characteristics of hydrogen lead to safety concerns which must be considered carefully in all stages of the solar-hydrogen economy.

Safety Considerations

Hydrogen combustion is not necessarily more dangerous than the combustion of natural gas or gasoline. It does however have differences which require different handling procedures to assure safety.

A major difference which may favor hydrogen use from a safety standpoint should be emphasized at the outset of this discussion. When hydrogen burns the combustion proceeds rapidly and dissipates upward rapidly. Hydrocarbons are heavier and they burn slower and closer to the ground. A gasoline fire in an open space would spread along the ground and burn for much longer than a hydrogen fire which would rapidly burn upward and out. On the other hand the more rapid combustion of hydrogen and its low fuel-to-air combustion ration make it more susceptible to explosive combustion in a closed space.

Another major difference which creates the need for different handling procedures is the fact that hydrogen burns with a colorless flame and is odorless. This makes it harder to detect than gasoline or natural gas to which an odor agent has been added. Of course, an odor agent could also be added to hydrogen.

Finally, the low ignition energy of hydrogen makes it necessary to assure that it is adequately contained and protected from electrical discharge and catalytic agents which might cause combustion.

Three factors are necessary for combustion to occur: an oxidizer, a fuel, and a source of ignition. It is generally accepted that two of these conditions must be controlled in order to assure complete safety when handling any volatile material. In the case of hydrocarbons however it is generally accepted that the complete control of the source of ignition is essential but that the separation of the fuel and the oxidizer is not. In the case of hydrogen this separation is considered essential.[24]

With proper system design this separation is not difficult to achieve. It requires that hydrogen be under sufficient pressure at all times during piping and storage to assure that it does not mix with air. This requirement is essentially the same as that for natural gas; however, the combustion characteristics of hydrogen, particularly its low fuel-to-air combustion ratio makes it particularly important. As with gasoline and natural gas, most safety procedures for handling hydrogen are associated with its storage and transportation. We will consider these procedures in the next chapter.

Undesired and unexpected combustion is of course only one of the safety hazards associated with fuels. Other hazards are toxicity and environmental pollution. Neither hydrogen nor methane are toxic to human beings. However, natural gas contains enough carbon monoxide to make it quite lethal in an enclosed space. Gasoline fumes are slightly toxic to humans.

When burned in air, all three of these fuels result in waste gases. In the case of the hydrocarbons some of these gases are quite harmful to both human beings and to the environment. The primary waste product in the combustion of hydrocarbons in air is carbon monoxide (CO). In addition to CO, nitrous oxide (NO), hydrocarbons (HC) and, when sulfur is present, sulfur dioxide, (SO_2) are also products. These pollutants also produce forms of photochemical oxidants when exposed to the ultraviolet light in sunlight. The oxidants produced—ozone, nitrogen dioxide (NO_2) and peroxyacetyl nitrate (PAN)—are the components of smog and are toxic to human beings. When they combine under the right circumstances their effect can be devastating. Four thousand people died in London in 1952 as a result of heavy smog which lasted for about four days.[25]

The carbon monoxide which results from hydrocarbon combustion combines with oxygen to make carbon dioxide. Carbon dioxide (CO_2) is used by green plants in the photosynthesis process, but far more CO_2 is produced in hydrocarbon burning than can be used by green plants. This excess CO_2 moves to the upper atmosphere where it collects and causes the greenhouse effect, which may be the most serious threat associated with the combustion of hydrocarbon fuels because of the changes in climate and weather patterns which it might cause.

The primary waste products created when hydrogen burns in air are water and nitrogen oxides (NO_x). However, since hydrogen burns over a wide range of temperatures the temperature of its combustion can be lowered to decrease the formation of NO_x or to eliminate it altogether and still retain a useful combustion temperature.

Domestic Residential Uses

The applications of hydrogen combustion for domestic residential uses encompasses all of those associated with natural gas, although minor adjustments in equipment are necessary.

When natural gas is used for domestic heating and cooking it is burned in a conventional Bunsen burner. In this type of burner the natural gas is mixed with air before it reaches the burner orifice.

Since the fuel/air ratio for hydrogen is so low and its flame speed is so rapid, if hydrogen were used in such a system it would burn before it reached the orifice. Such systems can be modified to use hydrogen by closing the air supply. Even though the temperature of the flame in such systems is less than that of natural gas, these systems will still suffice for cooking and heating purposes. However, there is noise generated due to flame flashback on both start-up and shutdown of modified burners.[26] As a result of the problems inherent in burners designed to use natural gas, it has been suggested that new burners designed only for hydrogen be installed on existing units when hydrogen becomes available for home use.

There are two basic approaches to hydrogen burner design which could easily be implemented. The first of these burner designs is similar to the Bunsen design except that the hydrogen is not mixed with air prior to combustion. In order to achieve adequately high flame temperatures the orifice of this type of burner must pass three times more hydrogen by volume than the Bunsen burner passes natural gas. While the design of this sort of system is feasible there appears to be another approach which offers greater potential.

When hydrogen and air combine in the pressure of a proper catalyst (platinum, for example) the hydrogen can be oxidized at very low temperatures. This is called *catalytic* combustion, and it can take place at room temperature depending on the activity level of the catalyst. By varying the type of catalyst used the temperature of the hydrogen combustion can be regulated. When the temperature is kept below 1500°F virtually no nitrogen oxides are produced. Catalytic burners could be produced and installed on home appliances and would not require pilot lights. Noble metal catalytic igniters could self-ignite when hydrogen is turned on and virtually no pollution would be created. More importantly, home heating systems using catalytic burners operating at below 1500°F could be built without the need for venting systems. Since the only product of the combustion would be a small amount of water vapor, the only requirement for their operation would be air and hydrogen. This possibility for home heating is very important when it is realized that as much as 30% of the chemical energy used in present-day furnaces is vented to the atmosphere. Research into catalytic combustion is being conducted at Billings Energy Research Corporation and at the Institute of Gas Technology (IGT)[27]

It can be seen that while certain adjustments to home heating and cooking systems would be necessary in order to use hydrogen, these adjustments could result in greater efficiency and less pollution than present-day uses of natural gas. An attempt to realize some of these benefits is being undertaken in the design and construction of domestic hydrogen use systems by the Billings Energy Research Corporation in Independence, Missouri.

The Billings group has constructed a hydrogen homestead which produces hydrogen by electrolysis, stores it in metal hydrides and uses it in gas appliances, in a heat pump and to fuel an auto.[28] This project has led to plans to convert a substantial portion of the town of Forest City, Iowa, to domestic hydrogen use. The hydrogen for this project would be produced in a large coal gasification to hydrogen plant located in Forest City.

Hydrogen combustion systems for home heating and cooking have important potential because they would result in less pollution and more efficiency as well as encourage the development of hydrogen production, transportation and storage systems. The use of hydrogen combustion systems in transportation vehicles offer similar potential.

Transportation Uses—Automobiles

Hydrogen combustion can be used in all three major transportation systems: automobiles, aircraft and railways. The first two of these systems have a long history of hydrogen application. The transportation system with which we are most familiar is the automobile powered by an internal combustion engine based on what is called the Otto cycle. The Otto cycle is a mathematical engineering model used to represent the ideal, theoretical relationship between gas pressure and volume in an enclosed chamber like the cylinder in an automobile engine. When the gas in this cylinder is compressed by a piston the volume decreases and the pressure increases. If after the gas is compressed, its pressure is caused to increase significantly and instantaneously by heat (theoretical) this increase in pressure will cause the volume to increase thus forcing the piston back down. This final increase in volume and the resulting movement of the piston converts the gas pressure into mechanical energy which turns the drive shaft and ultimately runs the automobile.

The gas used in a standard automobile is a mixture of air and gasoline vapors mixed in the carburetor. After it is compressed this mixture is forced to expand rapidly by the introduction of a spark from a spark plug which ignites it. Any gas which has the capability of being ignited under pressure will theoretically operate a standard internal combustion engine. It is relevant to note in this regard that the first internal combustion engine developed by Otto in the 1860s used gaseous fuels including hydrogen. Otto is said to have tried using gasoline in early experiments but the results were so "disastrous" that he gave it up and returned to the use of gas fuels.[29] Over the years we have become so accustomed to thinking of automobiles and gasoline together that the idea that an automobile can run on any other type of fuel seems almost unthinkable. Nevertheless, internal combustion engines do work well on other fuels, sometimes with greater efficiency and safety. The history of hydrogen use in internal combustion engines describes a technology that has developed a high degree of reliability over time, although it has not come into common use. As early as 1820 in a paper for the Cambridge Philosophical Society, Reverend W. Cecil suggested the use of hydrogen in a vacuum-type engine. We have noted that Otto himself used hydrogen as one of his fuels in early work on the internal combustion engine. However, interest in hydrogen as an engine fuel lagged as gasoline became more plentiful. The internal combustion engine designed to burn gasoline became the dominant technology and it was not until the 1920s that an interest in hydrogen as an engine fuel was revived.*

An early evaluation of hydrogen as a fuel was conducted during this time period by H. F. Ricardo, one of the primary developers of the internal combustion engine. Ricardo completed serious research on the burn characteristics of various hydrogen-to-air mixtures in engines.[30]

In the mid 1930s the German Government exercised considerable

* For a discussion covering the early history of the use of hydrogen in internal combustion engines see Roger E. Billings and Frank E. Lynch, *History of Hydrogen-Fueled Internal Combustion Engines,* (Independence, Missouri: The Billings Energy Research Corporation, Publication Number 73001, undated).

effort toward energy self-sufficiency, and hydrogen was the most logical and independent replacement for gasoline. Rudolph Erren was the person most responsible for the development of hydrogen as an engine fuel during this period. Erren worked in both England and Germany prior to World War II. He remained in England during the war but was interned by the British Government as an enemy alien and sent to Canada. He maintained a curious relationship with the English and German governments prior to the war and did considerable work on hydrogen engines for both. Erren developed a wide variety of hydrogen engine designs including submarine torpedos, aircraft and railcars. Unfortunately, most of his research records were lost during the bombing of Germany during World War II.[31]

In 1933 a paper by Erren and Hostings titled "Hydrogen: A Commercial Fuel for Internal Combustion Engines and Other Purposes" stimulated the interest of R. O. King who objected to some of the author's findings. King and his colleagues at the University of Toronto continued their research after the war and determined that the problem of backfiring in carbureted hydrogen/air engines was associated with high-temperature particles in residual gases and could be controlled with proper cleaning.[32] King was one of the few who continued to work on the hydrogen engine when cheap gasoline flooded the market at the end of World War II.

The 1960s brought a renewed interest in hydrogen as a fuel for internal combustion engines. In his early teens Roger Billings became interested in the possibilities of using hydrogen as a fuel. In 1966 at the age of sixteen he modified a model-A Ford truck to burn hydrogen and was on the way to becoming one of the foremost hydrogen use proponents and developers in the United States. Billings has since converted several late-model automobiles to hydrogen use and completed a number of projects demonstrating the feasibility of hydrogen as a fuel for buses and for mail trucks.[33] Billings has helped to pioneer the overall effort to convert existing internal combustion engines for hydrogen use. Among his major contributions is the injection of water vapor into the burning mixture which reduces the heat of combustion and virtually eliminates the formation of nitrogen oxides.[34] The process also helps control pre-ignition and flashback.

In the 1970s there was a strong upsurge of interest in hydrogen as a fuel for automobiles and buses. This interest was spurred by the gasoline shortages and the energy dependency which these shortages highlighted. Daimler-Benz displayed a hydrogen-powered van at the 1975 Frankfurt Auto Show. This van used a metal hydride hydrogen storage system and was similar in appearance to the standard Daimler-Benz van. Since this effort Daimler-Benz has developed a hydrogen-powered passenger car and a city bus.[35]

Interest in the hydrogen-fueled engine continues to increase. In 1978 the Soviet Union reported that Soviet scientists had developed a Volga automobile to run on both gasoline and hydrogen and that taxis and a bus would be street-tested that year. In 1979 the Musahi Institute of Technology in Japan displayed a compact hydrogen-powered auto using liquid hydrogen which had excellent range, speed and operating characteristics.

While interest in hydrogen as a fuel for internal combustion engines has never died, it has fluctuated over time and this fluctuation seems to be a function of the availability of gasoline.

A brief comparison of the use of hydrogen and gasoline in internal combustion engines however demonstrates that hydrogen has benefits which go well beyond those associated with its value as a substitute fuel.

As we noted earlier hydrogen has more than twice the heating value by weight than that of the hydrocarbons. However, by volume it contains less energy. This energy differential is due to the lesser density of hydrogen. Hydrogen also burns over a rather wide range of fuel-to-air mixtures. A mixture of hydrogen and air will burn with as little as 4% by volume of hydrogen present. On the other hand such a mixture will also burn with as much as 75% hydrogen. The stoichiometric mixture, we have noted, is that mixture which yields the highest possible temperature of burning. For hydrogen this is a mixture composed of about 30% hydrogen.

We also noted that because hydrogen burns very rapidly, its flame speed is almost 10 times more rapid than hydrocarbons. We noted further that the energy necessary for ignition of hydrogen is quite low. However, the auto ignition temperature for hydrogen is almost 40% higher than that for gasoline. This higher temperature means that greater pressure can be achieved before hydrogen will ignite without an outside source.

Gasoline is a very different sort of fuel. Though highly volatile, it is a liquid at standard temperature and pressure. It vaporizes easily and the vapor tends to remain close to the source rather than dissipating rapidly.

Gasoline burns over a narrow range of fuel/air mixtures. About 1.3% of gasoline by volume with air is necessary for combustion, and mixtures containing much more than 7% of gasoline are too rich for combustion. The stoichiometric mixture for gasoline and air is about 1.7%, which is relatively close to the lower flammability limit.

Gasoline has a relatively slow flame speed and higher ignition temperature than hydrogen.[36] However as we noted it has almost a 40% lower auto ignition temperature.

When the differences noted above are compared with regard to the use of these two fuels in internal combustion engines some very important observations are possible. Hydrogen and gasoline use in internal combustion engines can be compared from three perspectives: efficiency of engine power generation, waste product formed on combustion, and storage and handling.

Since hydrogen is so much less dense than gasoline a larger volume of it must be carried in order to achieve the same energy transfer when hydrogen is used in a converted gasoline engine. On the other hand the auto ignition temperature of hydrogen is a bit over a third greater than that of gasoline. Theoretically this higher ignition temperature should allow for the design of hydrogen-burning engines with a third greater thermal efficiency which could help to offset the volume-ratio problems.

The fact that hydrogen is so much less dense than gasoline means that the volume ratio of hydrogen to air will have to be greater than that for gasoline and air. This condition is reflected in the difference in the stoichiometric ratios of the two.

For maximum energy release on combustion the volume of hydrogen to air must be almost 20 times greater than that of gasoline and air. In internal combustion engines with similar cylinder size this difference in fuel/air ratio means that the engine using hydrogen operates at about 66% the power output of the gasoline-fueled engine. This condition holds for all engine designs which mix the fuel and air prior to introducing the fuel into the cylinder. Since the volume of air is the controlling factor, when hydrogen is injected under pressure into a cylinder after the air has been compressed the problem is overcome. In this circumstance the hydrogen combustion engine will have greater power output because the hydrogen and air mixture releases more energy than the gasoline and air mixture when the same volume of air is consumed.

It is clear from this examination that hydrogen-air internal combustion is potentially more efficient in terms of power output due its greater energy content by weight and the fact that a mixture of hydrogen and air can exist at much higer pressure than can a mixture of gasoline and air. However, in order to realize this greater potential efficiency the standard internal combustion engine would have to be modified to take advantage of these characteristics. This modification would have to include the injection of hydrogen to the cylinder after the air intake valve is closed and methods for increasing the pressure of air in the cylinder during the compression phase. Even though the hydrogen air engine could be more powerful than the gasoline engine it would still not be able to equal the gasoline auto's fuel-storage-size-to-power ratio because the weight of an equal volume of liquid hydrogen is less than that of gasoline. Therefore, more volume for fuel storage must be provided in hydrogen vehicles to obtain the same weight of fuel as a gasoline powered vehicle. This problem is further compounded when the weight of liquid hydrogen storage tanks is taken into consideration.

The technical problems involved in modifying a standard gasoline engine for hydrogen can be considered in two categories: those involved in converting the engine from liquid fuel to gas, and those resulting from the particular nature of hydrogen. The first of these problems is easily handled through the use of a carburetor designed to mix air and another gas. This is essentially a similar solution whether the gas is methane, propane, or any other gas fuel. Once this modification is made, however, there are some particular characteristics of hydrogen combustion which necessitate further modification. We have noted that hydrogen mixes rapidly with air and that it burns over a wide range of air-to-fuel ratios. Due to these characteristics the carburetor of a hydrogen-burning engine as well as the valve system must be designed to prevent the hydrogen and air mixture from burning in the carburetor or the fuel/air intake system. This is called *flashback*.

The low energy necessary to ignite hydrogen and air can result in a phenomenon called *pre-ignition.* This condition results when the fuel air mixture ignites prior to the desired time. Van Vost and Woolley note that both flashback and pre-ignition were recognized as problems by a number of early hydrogen combustion researchers. R. O. King suggested that pre-ignition was caused by hot carbon on the cylinder walls and pistons, and that it could be overcome by keeping the engine clean. Roger Billings suggested that it was caused by a premature spark due to voltage leakage.[37]

These problems can be solved by reducing the combustion temperature of the hydrogen and air mixture either by recirculating cooled exhaust or by injecting water vapor into the mixture. The latter method was used by Billings. Both of these methods of reducing the combustion temperature appear to reduce the temperature of hot spots on the cylinder and piston thereby reducing pre-ignition. Both of these techniques also result in a proportional reduction in power. The problems of pre-ignition and flashback can be solved more fundamentally by injecting the hydrogen into the cylinder after air compression has taken place. This would assure that the combustion mixture was present only where it was intended to burn.[38]

In recent years people have become increasingly more aware of the hazardous effects of the waste products of combustion. The waste products which result from combustion often depend on the temperature at which the combustion occurs. We have noted that when gasoline and other hydrocarbons burn they produce carbon monoxide, oxides of nitrogen and other hydrocarbon compounds. The carbon monoxide is further converted to carbon dioxide which collects in the upper atmosphere and creates the greenhouse effect. Oxides of nitrogen, when exposed to sunlight, are converted to the complex chemical mixture which is smog. They also migrate to the upper atmosphere and cause a deterioration of the ozone layer. Carbon monoxide and hydrocarbons are composed of the chemical components of the hydrocarbon themselves but the nitrogen oxides are a result of the combustion of nitrogen in air with oxygen. This reaction takes place as a result of the heat produced during combustion. The maximum heat of combustion of gasoline takes place at a gasoline-to-air mixture ratio (the stoichiometric mixture) which is very close to the mixture ratio necessary to support combustion. The percent of gasoline in air necessary for combustion to occur is about 0.4% less than that which will provide the greatest heat of combustion. Both the temperature of combustion of the lowest combustion ratio and that for the stoichiometric ratio are sufficiently high to cause the formation of nitrogen oxides.

Hydrogen will also cause nitrogen oxide to form when it is burned in air at high enough temperatures. However, as we have noted hydrogen can be made to burn at a temperature below that which is necessary for the formation of oxides of nitrogen. There are two major ways of lowering the temperature of hydrogen combustion. The first of these is to reduce the fuel/air mixture ratio. Whereas the difference between the mixture ratio necessary to support combustion and the ratio resulting in the highest temperature for gasoline is only 0.4%, it is 25% for hydrogen. This high range of burning mixtures for hydrogen means that the mixture ratio of hydrogen to air can be reduced considerably with a corresponding reduction in combustion temperature. This ratio can be reduced enough to prevent the formation of nitrogen oxides while still providing enough energy to adequately power an internal combustion engine. Under these conditions the only waste product of a hydrogen/air internal combustion engine would be water. Another way of reducing the temperature of combustion of hydrogen and air is to inject water into the mixture. We noted that this process also controls pre-ignition and flashback.

Storage and handling of hydrogen for use in automobile engines are different than the storage and handling of gasoline. The objects of any given fuel storage and handling system are to provide optimum fuel, with the least volume and weight, and the greatest safety and reliability. To this last must be added considerations of convenience and cost. While all of these conditions may seem at first glance to be separate concerns they are completely interrelated. Consider, for example, the concept of optimum fuel capacity of a vehicle. The amount of fuel that a vehicle will be required to carry will depend on the distance between fueling stations along the desired route of the vehicle. Therefore the size of the tank must be a function of the network of fuel centers. The size of the tank can be reduced if the load it carries is reduced or the fuel is made to burn more efficiently. All factors must be judged according to safety and reliability, and the price and convenience of the systems must be weighed against the costs of using it.

Gasoline is easier to contain than hydrogen and it also has a better power-to-volume ratio. It exists as a liquid at room temperature, can be pumped as a liquid and can remain in the storage tank for relatively long periods. Since the use of gasoline is so widespread and the network of distribution so intricate the size of a vehicle storage tank needs to be of minimum concern. Gasoline storage and handling systems are mechanically reliable and reasonably safe.

Hydrogen is a gas at room temperature and is more difficult to contain as a gas or a cooled liquid. On the other hand hydrogen can be stored in metal hydrides conveniently and much more safely than liquid gasoline. In any case the weight and/or volume required for hydrogen storage on board a vehicle is a serious consideration. The lack of a network for the wide distribution of hydrogen fuel is a great limitation which at this point cannot be overcome by increasing the size of tanks.

This brief comparison of the hydrogen and gasoline as fuels in internal combustion engines indicates that the two compare favorably with regard to power and efficiency. Standard gasoline engines which are modified to burn hydrogen are less efficient and powerful. However, engines designed and built specifically for hydrogen promise to be more efficient and powerful than gasoline engines. The combustion characteristics of hydrogen make it almost totally nonpolluting whereas gasoline creates dangerous waste products on combustion. Finally, hydrogen is more difficult to store than gasoline which also has a better volume and weight to available energy ratio.

Transportation Uses—Aircraft

The potential use of hydrogen as a fuel for aircraft began to excite the imagination of some farsighted individuals shortly after World War II. As early as 1918 hydrogen had been considered and rejected as an aviation fuel in piston engines. However, with the development of the jet engine hydrogen became much more attractive. Hydrogen burns smoothly in turbo jet engines. These engines are essentially more simple than piston engines and problems of pre-ignition and flashback are not inherent difficulties.

The fuel commonly used in jet engines is a kerosene called Jet A. It has a heat of combustion of 18,400 Btu/lb. Hydrogen has a heat of combustion of 51,590 Btu/lb. As a result of this difference, nearly three times as much Jet A by weight is required to deliver the same amount of power as hydrogen.[39] The heat of combustion advantage of hydrogen was recognized early in the history of jet aviation and was responsible for much of the early interest. There were other conditions which made hydrogen attractive, however, and these had to do with the new demands placed on aircraft. After World War II there was great interest in the United States military establishment in making aircraft go higher, longer, and further, as John Sloop notes in his 1945–1959 history of hydrogen fuel. Hydrogen had a theoretical advantage of operating an aircraft at higher altitudes due to its low air-to-fuel combustion ratio. Additionally, at high altitudes there is less friction due to air (*drag*) because of the lower air pressure. Airplanes could then be made larger to accommodate the increased volume of hydrogen necessary to take advantage of the heat of combustion benefits. At lower altitudes such an increase in aircraft size was considered a serious disadvantage. Aircraft using hydrogen fuel would be larger than conventional planes because of the greater volume of fuel necessary to provide an equivalent weight of Jet A. However, they would weigh the same and fly higher and farther. In 1954 these potential advantages caught the imagination of Abe Silverstein, the Associate Director of Lewis Laboratory of the National Advisory Committee for Aeronautics (NACA). Silverstein's interests coincided with those of the air force, and a secret project to determine the feasibility of using hydrogen as an aircraft fuel was begun in 1955. This effort, called Project Bee, successfully used liquid hydrogen to power a B-578 twin engine bomber using a Curtis Wright J-65 turbo-jet engine. The first successful flight took place in February 1957.*

Work on the hydrogen-powered aircraft parallelled the efforts to develop a new high-altitude reconnaissance aircraft. It seemed clear that hydrogen could be used to good advantage in such an aircraft if for no other reason than that it would burn consistently at high altitudes. The Lockheed U-2 high-altitude spy plane in service since 1956 used conventional jet fuel which would not re-ignite at high altitudes.

---

* Most of the information in this section on the history of hydrogen as an aviation fuel is extracted from an outstanding history of the subject authored by John L. Sloop in 1978. It is unfortunate that this publication has not received wider distribution. See John L. Sloop, *Liquid Hydrogen as a Propulsion Fuel, 1945–1959,* (Washington, D.C.: Nasa SP4404, 1978).

If a flameout occurred it had to decend to 9000 meters in order to relight. Many experts felt that exactly that had occurred when the U-2 piloted by Francis Gary Powers was shot down over the USSR in 1960.

After an aborted effort to develop a project based on a new hydrogen aircraft engine called the REX designed by R. S. Rae, the airforce instituted a super secret project to develop a hydrogen-powered reconnaissance aircraft. This project, called operation Suntan, was so secret that full access was limited to only 25 people and even now few people are aware that it ever happened.[40] The project was designed to develop a successor to the U-2. It resulted from a proposal to the airforce submitted by Clarence L. (Kelly) Johnson, Chief Engineer of Lockheed Aircraft and the father of the U-2. The airplane to be built was called the CL-400. It was to have a speed of Mach 2.5, fly at an altitude of 30,300 meters, and a range of 4070 kilometers. The engine for this aircraft was built by Pratt and Whitney, called the Model 304. The development of this engine was proceeding satisfactorily but had been test-operated for only a total of 25.5 hours when the project ceased.

No clear reasons for the failure of this project are available even though between 100 and 250 million dollars were spent. Kelly Johnson seems to have lost some confidence in the project in its early phases largely as a result of a disagreement with the Air Force on the design range of the aircraft. However, it appears to have been General Curtis E. LeMay who as Vice Chief of Staff of the Air Force really doomed the project. LeMay seems to have been convinced that there were international political risks involved with the use of hydrogen. What is probably a more direct reason for his lack of support for the project was his support of a competing project toward which he could divert the Suntan funds. In any case the project folded in February, 1959.

While this project advanced hydrogen fuel technology considerably, it did little to convince the government of the potential of hydrogen-fueled aircraft. The reason for this failure is probably twofold. First, the project itself seems to have been designed around an immediate military goal, that of developing an improved spy plane. In order to be successful such a plane might have to operate from remote bases where supplies of hydrogen might be hard to maintain.

Secondly, perhaps again a result of the project's limited objectives, the full benefits of hydrogen fuel were not taken into account. No mention was ever made of the environmental benefits and apparently no thought was given to the potential scarcity of hydrocarbon fuel. Given these two reasons for failure, the first of which was certainly clear at the outset of the project, it is surprising that the project was ever undertaken at all.

Advances in hydrogen fuel technology which developed out of project Suntan were directly applied to NASA's space vehicle propulsion program. The use of liquid hydrogen and oxygen to fuel all of the United States' successful space flights is a testimony to the sophistication of current technology.

Since the failure of project Suntan there have been several expressions of interest in hydrogen-powered aircraft but Lockheed alone seems to be promoting hydrogen for subsonic flight. In 1978 Lockheed proposed a project called LHEAP (Liquid Hydrogen Experimental Airline Project). This small, four-plane cargo project would link the U.S., Europe and the Middle East. It would be the first continuous aircraft operation using hydrogen fuel and would employ a converted Lockheed L-1011 Tri-Star wide body jet. Thus far, Lockheed has not been successful in creating a great deal of interest in the project.[41]

The use of hydrogen as a fuel for aircraft has clear advantage over Jet A type fuel. Hydrogen is lighter by about a factor of three for equal power; it can be made to burn with virtually no pollution; it is a universal fuel which can be made available anywhere; and it can, as we have seen, be made from all energy sources. In addition the technology for hydrogen flight is available now as a result of considerable technological development over the last forty years.

NOTES

1. Bernard J. Crowe, *Fuel Cells, A Survey,* (Washington, D.C.: National Aeronautics and Space Administration, 1973), pp. 27–29.
2. Ibid., p. 37.
3. Ibid.
4. G. J. Young, *Fuel Cells,* Vol. II, (New York, Reinhold Publishing Corporation, 1963), pp. 2–3.
5. Ibid.

6. Crowe, p. 74.
7. Crowe, p. 76.
8. A. S. Wilk, *Plan for the Development of Fuel Processors for Fuel Cell Systems,* (Washington, D.C.: The Engineering Societies Commission on Energy, Inc., 1978).
9. Wilk, p. 31.
10. W. R. Nixon et al., *Market Assessment of Fuel Cell Total Energy Systems Summary Report,* (Oak Ridge, PA: Oak Ridge National Laboratory, 1979).
11. J. O'M Bockris, *Energy: The Solar-Hydrogen Alternative,* (New York: Halsted Press, 1975).
12. Crowe, pp. 80–84.
13. Crowe, p. 85.
14. Peter Hoffmann, *The Forever Fuel, The Story of Hydrogen;* Boulder: Westview Press, 1981 p. 136
15. Crowe, p. 50.
16. For example see E. N. Balko and W. Avery, "*Fuel Cell Efficiency in a Combined Electrochemical-Thermochemical Process,*" *International Journal of Hydrogen Energy,* (Great Britain). Vol. 6. No. 3. 1981, pp. 255–266.
17. Bokris, p. 214.
18. Amory Lovins. *Soft Energy Paths,* (New York: Harper and Row, 1977), pp. 88–89.
19. *Distributed Energy Systems: A Review of Related Technologies,* (Washington, D.C.: Arthur D. Little, Inc for U.S. Department of Energy, 1979), DOE/PE 03871-01, pp. 16–19.
20. Ibid.
21. Crowe, p. 62.
22. Archie W. Culp, Jr., *Principles of Energy Conversion,* New York: McGraw-Hill Book Company, 1979), pp. 399–400.
23. J. Pangborn and M. I. Scott, "Domestic Uses of Hydrogen" *Hydrogen: Its Technology and Implications,* Vol. IV, ed. Kenneth E. Cox and K. D. Williamson, Jr., CRC Press, Cleveland, Ohio, p. 172.
24. F. J. Edeskuty, "Safety", *Hydrogen: Its Technology and Implications,* Vol. IV, ed. Kenneth E. Cox and K. D. Williamson, Jr., CRC Press, Cleveland, Ohio, p. 205.
25. J. O'M Bockris. *Energy Options, Real Economics and the Solar-Hydrogen System,* (New York: Halsted Press, 1980), p. 288.
26. Pangborn, p. 179.
27. Pangborn, pp. 184–186.
28. Roger E. Billings, "Hydrogen Homestead," (Presented at the 2nd World Hydrogen Energy Conference, Zurich, Switzerland), 1978.
29. W. D. VanVost and R. L. Woolley, "Hydrogen Fueled Surface Transportation," *Hydrogen: Its Technology and Implications,* Vol. IV, ed. Kenneth E. Cox and K. D. Williamson, Jr., 1979, p. 4.
30. W. D. VanVost and R. L. Woolley, p. 5.
31. Hoffmann, pp. 30–31.

32. Billings.
33. Hoffmann, p. 112.
34. Ronald L. Woolley, "Hydrogen Engine NOx Control by Water Induction," A paper submitted to *NATO/CCMS Fourth International Symposium on Automotive Propulsion Systems,* April 1977, (The Billings Energy Corporation, February 1977).
35. Hoffmann, pp. 123–124.
36. For a good comparison of the relative burning properties of hydrogen and gasoline see W. D. Van Vost and R. L. Woolley, *Hydrogen Fueled Surface Transportation,* pp. 8–10.
37. Van Vost, p. 12.
38. Van Vost, pp. 13–14.
39. G. D. Brewer, "Hydrogen-Fueled Aircraft," *Hydrogen: Its Technology and Implications,* Vol. IV, eds. Kenneth E. Cox and K. D. Williamson, Jr., CRC Press, Cleveland, Ohio, 1979, p. 89.
40. John L. Sloop, *Liquid Hydrogen as a Propulsion Fuel, 1945–1959,* Washington, D.C., NASA SP4404, 1978, p. 143.

REFERENCES

Balko, E. N. and Avery, W. "Fuel Cell Efficiency in a Combined Electrochemical-Thermochemical Process." *Int. J. Hydrogen Energy.* Great Britain. Vol. VI. No. 3. 1981.

Billings, Roger E., *Hydrogen Homestead.* Presented at the 2nd World Hydrogen Energy Conference. Zurich, Switzerland. 1978.

Billings, Roger E. and Frank E. Lynch. *History of Hydrogen-Fueled Internal Combustion Engines.* Publication Number 73001, Independence, Missouri: The Billings Energy Corporation, undated.

Bockris, J. O'M. *Energy Options, Real Economics and the Solar-Hydrogen System.* New York: Halsted Press, 1980.

Bockris, J. O'M. *Energy: The Solar-Hydrogen Alternative.* New York: Halsted Press, 1975.

Brewer, G. D. "Hydrogen-Fueled Aircraft. *Hydrogen: Its Technology and Implications.* eds. Kenneth E. Cox and K. D. Williamson, Jr., CRC Press, Cleveland, Ohio. Vol. IV, 1979.

Crowe, Bernard J. *Fuel Cells, A Survey.* Washington, D.C.: National Aeronautics and Space Administration, 1973.

Culp, Archie W., Jr. *Principles of Energy Conversion.* New York: McGraw-Hill Book Company, 1979.

Distributed Energy Systems: A Review of Related Technologies. Arthur D. Little, Inc. for U. S. Department of Energy, DOE/PE 03871-01, 1979.

Edeskuty, F. J. "Safety" *Hydrogen: Its Technology and Implications.* eds. Kenneth E. Cox and K. D. Williamson, Jr., CRC Press, Cleveland, Ohio, Vol. IV.

Nixon, W. R. et al., *Market Assessment of Fuel Cell Total Energy Systems Summary Report.* Oak Ridge, PA: Oak Ridge National Laboratory, 1979.

Pangborn, J. and Scott, M. I. "Domestic Uses of Hydrogen" *Hydrogen: Its Technology and Implications.* eds. Kenneth E. Cox and K. D. Williamson, Jr., CRC, Press, Cleveland, Ohio, Vol. IV.

Roney, James R. "OTEC to Hydrogen Fuel Cells—A Solar Energy Breakthrough." *Hydrogen Energy Progress, The Proceedings of the 3rd World Hydrogen Energy Conference.* Tokyo, Japan: Vol. III, June, 1980.

Sloop, John L. *Liquid Hydrogen as a Propulsion Fuel 1945–1959.* Washington, D.C.: NASA SP 4404, 1978.

Van Vost, W. D. and R. L. Woolley. "Hydrogen Fueled Surface Transportation." *Hydrogen: Its Technology and Implications,* eds. Kenneth E. Cox and K. D. Williamson, CRC, Press Cleveland, Ohio, Vol. IV, 1979.

Wilk, A. S. *Plan for the Development of Fuel Processors for Fuel Cell Systems.* Washington, D.C.: The Engineering Societies Commission on Energy, Inc., 1978.

Woolley, Ronald L. *Hydrogen Engine NOx Control by Water Induction.* A paper submitted to *NATO/CCMS 4th International Symposium on Automotive Propulsion Systems,* April, 1977. The Billings Energy Corporation, February, 1977.

Young, G. J. *Fuel Cells,* Vol. II. New York: Reinhold Publishing Corporation, 1963.

6. STORAGE AND TRANSPORTATION OF HYDROGEN

Ironically, the aspects of hydrogen's greatest potential as an energy common denominator present the greatest technological challenges. From an energy efficiency perspective hydrogen is the best vehicle for storing and transporting energy. Almost any source of energy can be converted into hydrogen, and hydrogen can be moved with much less energy loss than electricity. However, there are difficulties associated with the storage and transportation of hydrogen which must be specifically confronted before these advantages can be realized. Many of these difficulties, particularly safety aspects, are related to a general lack of familiarity with hydrogen and are no greater than those associated with the storage and transportation of gasoline or natural gas. However, some of these difficulties are a result of the unique characteristics of hydrogen.

Like natural gas hydrogen in its common form is lighter than air and highly combustible. Also like natural gas it must be stored and piped under pressure. This storage and transmission is not an insurmountable problem and the technology for doing it has been developed to a high degree of reliability by the natural gas industry. Two characteristics of hydrogen, however, complicate the problem of storage and transportation. The first of these is hydrogen's low density. As a result of its low density more care must be taken to assure that pipes and tanks are leakproof. The second, and more serious, problem is hydrogen's high chemical activity which, under certain circumstances, can result in hydrogen *embrittlement* of metal pipes and storage containers. Embrittlement is a function of the nature of the metal, the pressure of storage, and the temperature. It is controllable but warrants design consideration in any hydrogen storage system.

Another aspect of the high chemical activity of hydrogen is the nature of hydrogen combustion. We have noted that hydrogen burns rapidly over a wide range of temperatures and air/fuel ratio mixtures. These characteristics make the safety consideration of hydrogen storage and transportation very important. Safety considerations are no less important in the storage and transportation of gasoline and natural gas. But natural gas and gasoline are more familiar, and procedures for their storage, transportation and use have been developed over time. Reliable and safe procedures have also been developed for handling hydrogen under certain conditions. These procedures have been used by NASA for years in conjunction with the use of liquid hydrogen in space vehicles. Liquid hydrogen is routinely transported by rail and frequently by truck with an admirable safety record and probably less potential danger than the shipment of some chemical stock.*

Safety considerations in the storage and transportation of hydrogen are important, and procedures for handling must be developed as carefully and responsibly as those for handling gasoline and natural gas. Some of the procedures for handling hydrogen safely differ from those for handling gasoline and natural gas. These differences are not a reflection of a necessarily greater inherent danger in the case of hydrogen. Rather, they stem from differences in chemical characteristics. We will discuss some of these safety concerns in more detail later.

In recent years some aspects of the storage and transportation of liquid hydrogen have reached a high state of safety and reliability due to the space program. Since liquid hydrogen is an important element in the transportation aspects of the solar-hydrogen economy we will consider these procedures. Other techniques for the storage and transportation of hydrogen however offer a wide range of possibilities for the whole system. The storage and transportation of hydro-

---

* The 1977 train derailment in Pensacola, Florida, which resulted in the rupture of a tank car filled with nitrous ammonia and the deaths of several persons as a result of the gas is a case in point. No similar disaster has happened in transporting hydrogen.

gen as a free gas holds great promise for domestic uses. Transportation (transmission) techniques are not only compatible with the existing natural gas system, but they also allow for the gradual introduction of hydrogen through mixing the gases in the same pipelines. We will discuss several aspects of free gas transmission in this chapter.

Its unique chemical nature gives hydrogen a range of storage possibilities which are available to no other fuel. Some of these possibilities are inherently safer and more convenient than any other energy storage system. These storage systems are in the early stages of development and promise great reliability, safety and flexibility. The most developed of these new storage systems is *metal hydride* storage. Hydrogen combines with some metals to form hydrides under specific conditions of temperature and pressure. These hydrides release hydrogen on the application of heat. Hydrogen can thus be stored as a metal hydride in an inert condition until needed and then released with the application of small amounts of heat. Another new development is hydrogen storage in *glass microballoons.* This system, similar to metal hydride storage, keeps hydrogen safely isolated from air. It also has the advantage of being a low weight system of storage. We will examine these and other new techniques in detail in this chapter.

While the storage and transportation of hydrogen are clearly technological and procedural challenges, answers to these challenges seem available and replete with possibilities for the development of fuel-handling systems which may be far safer and more convenient than those we now have.

We will consider three major types of hydrogen storage systems in this section: pressure storage, metal hydride storage and liquid storage. In the category of pressure storage we include such conventional technologies as tank or pipe storage and underground storage as well as newer technologies such as glass microballoon storage. Metal hydride storage is the bonding of the hydrogen molecule with the molecular structure of certain metal alloys. Finally, liquid storage refers to hydrogen which has been cooled under pressure until it takes a liquid form.

PRESSURE STORAGE

Pressure storage of hydrogen can be considered in three major technological categories: pressure vessel storage, microballoon storage, and underground storage. While somewhat arbitrary these categories allow us to focus on related construction and safety problems.

There are several types of pressure vessels available for storing hydrogen. Perhaps the most familiar is the common gas-cylinder bottle similar to that used for oxygen. This type of vessel illustrates one of the problems associated with hydrogen storage. The low density of hydrogen gas creates a poor volume/energy ratio. In order to improve this ratio it is necessary to store hydrogen under pressure, but vessels with the necessary strength to sustain the pressure, such as gas cylinders, are often heavy and expensive. A way to overcome part of the expense problem is to use materials in common production. One such material is seamless steel pipe. Storage of hydrogen in banks of capped pipes is one of the more attractive methods for pressure vessel storage. It must be emphasized that the economic advantage of this system over cylinder storage is that pipe material is commonly produced for other purposes. The problem posed by the high volume/energy ratio is not completely solved by storage in pipe banks, nor is the more serious safety problem caused by hydrogen embrittlement.

This problem of embrittlement is a result of the tendency of the hydrogen molecule to react with the metal of the container. The reaction of hydrogen molecules and the molecular structure of the metal container is quite complex and it involves many variables. The overall effect is a drastic reduction of the durability and strength of the material. Embrittlement takes place over a wide range of temperatures but the process changes significantly at temperatures above 440°F. Of most direct concern is the reaction which takes place at ambient temperature. There are two general types of reactions which occur: instantaneous reaction where no hydrogen has diffused into the lattice structure of the metal, and delayed reaction which results from the progressive action of hydrogen in lattice of the metal.[1]

Embrittlement appears to be a function of at least three conditions in addition to heat. First, the nature of the metal affects the degree of embrittlement. In general, embrittlement in steel alloys tends to increase with the strength of the metal but other alloy characteristics also are important. A second condition is the pressure of the stored hydrogen. As the pressure of the gas is increased the severity of embrittlement increases depending on the alloy. Finally, embrittlement increases with the purity of the stored hydrogen.

In high-temperature systems involving steel containers embrittlement also involves surface decarbonization as well as hydrogen diffusion into the lattice structure of the metal. Both of these conditions can be controlled by adding carbide-stabilizing elements like chromium to the steel.[2] At the present time hydrogen is routinely stored in the tube banks at pressure of about 2000 psi. However, higher pressure storage, up to 15,000 psi, has been safely attained and used by NASA in the space program.[3]

Of the steel containers, the forged, low-alloy, storage containers and storage vessels made of low-strength, structural-grade steel appear to be unaffected by hydrogen embrittlement. Alloys of aluminum also appear to offer possibilities. One such alloy was used in the pressure vessels in the Mariner 1969 mission to Mars. Large-scale storage of hydrogen however may best be accomplished through the use of nonmetalic containers.

Pressure vessels constructed of concrete, and vessels wrapped with high-strength fiberglass (filament wound) have also been considered.[4] Prestressed concrete vessels with hydrogen-impermeable liners may have important advantages, particularly when used for large-scale storage and buried underground to take advantage of the overburden pressure of the earth. Filament-wound vessels are composed of light, hydrogen-impermeable materials wrapped with high-strength fiberglass. While these vessels show no present cost advantage they do offer possibilities for use where mobile storage may be necessary.[5]

Another important potential technology for the pressure storage of hydrogen is microstorage either in glass microballoons or hollow fibers. Microballoon storage is the pressure storage of hydrogen in tiny glass spheres of 20 to 200 microns in diameter. A handful of these spheres is similar in appearance to powdered sugar. The spheres are derived from fly ash. In order to charge or fill these spheres with hydrogen, they are heated to about 1200°K or the temperature necessary to decrease the gas permeability of the glass. Hydrogen is then introduced at the necessary pressure and the balloons are cooled so that their permeability decreases. The hydrogen which is trapped in these small glass balloons will remain until they are again heated to a temperature which will allow the gas to escape.[6] While there is much yet to learn about this system it appears to offer the potentials of stability, mobility, lightness, and above all, safety. Such a system might have applicability for hydrogen storage in passenger vehicles and home storage of hydrogen.

The storage of hydrogen in natural or constructed underground cavities can be considered a form of pressure storage on a very large scale. The technology for the storage of gas underground is relatively old, dating from the early 1900s. By the 1950s underground cavities were used to store propane and other light gases such as helium. While hydrogen itself has not yet been stored underground in the United States it has been successfully stored underground in the United Kingdom in salt caverns.[7]

There are three essential conditions which must be met for the successful underground storage of any gas: adequate gas containment characteristics of the site, adequate gas capacity to meet the storage need, and the deliverability of the site. The containment characteristics depend on the geological geometry of the site which must include as a minimum an impervious caprock, or upper roof, usually in the shape of a shallow dome, covered with an overburden of sufficient weight to contain the gas under pressure. Other geometric considerations are the depth of the storage area, the thickness, the nature of geological faults in the area, and the overall shape of the enclosure. The capacity of a reservoir depends on the containment characteristics as well as the reservoir pore volume (a function of porosity and bulk volume), reservoir pressure and temperature.

The deliverability of a reservoir is the capability of the site to yield the gas on demand at a high enough rate for the market demand. It is dependent primarily on the permeability of the reservoir, the pressure level, and the number and spacing of wells.

There are three major types of underground storage sites for hydrogen available in the United States: depleted gas and oil reservoirs, aquifers, and engineered caverns such as salt and coal mines. In addition there is the potential of developing underwater sites.

The feasibility for developing underground hydrogen storage facilities appears to be quite good; however, there are several important concerns which must be considered. These concerns involve questions of storage technology, economics, site location, and safety.[8]

The technology of underground gas storage is quite complex and the process itself varies with geological conditions. Development of any storage site for underground gas storage involves the careful exploration and evaluation of the site with regard to the geological features mentioned earlier. In the case of hydrogen storage additional considerations are required. Evaluations must be made regarding the chemical reactivity of hydrogen with the surrounding materials. This consideration is important in the case of depleted gas and oil wells where hydrogeneration is a possible reaction. Another concern is related to the tendency of hydrogen to go into solution and diffuse in ground water. This is a problem with hydrogen storage in aquifer reservoirs.[9]

The growth of an underground gas storage system, particularly one for the storage of hydrogen, is a dynamic process which involves the development of a gas storage bubble in the reservoir. As this bubble develops it migrates to the upper sections of the reservoir. Once the process of bubble formation stabilizes, lower pressure tends to exist around the outer edges and some of the gas goes into a buffer or cushion status, an apparently useful but unrecoverable state.[10]

The technical problems associated with hydrogen storage underground appear to be reasonably manageable depending on the storage site. Experience indicates that storage in depleted salt mines or limestone caverns would pose the least technical challenge. However, the problems associated with aquifer storage and depleted hydrocarbon well storage also appear to be solvable.

Particular safety problems associated with the underground storage of hydrogen relate more or less directly to the wide range of temperatures over which hydrogen burns and its diffusability which makes it susceptible to leakage. These problems require that any underground storage system for hydrogen be carefully monitored and that special fail-safe systems be installed.

Site selection and economics of underground hydrogen storage are to a large extent dependent on the deliverability and distance of transmission to desired markets. In general there is a wide distribution of potential underground storage sites over the United States. M. R. Tek estimates that there are over 6000 gas reservoirs, 132,000 gas wells and 70,000 linear miles of gathering networks in the 48 continental states. Many of these he notes are in some state of depletion and some are ready for use.[11] In addition, there are numerous mine cavities that are available in the gulf states, southeastern coastal states and the western states. It would appear that there is more than adequate potential underground storage available to sustain a hydrogen economy in the United States.

Any economic analysis of the potential of underground storage of hydrogen would at this point by highly speculative and probably misleading as a result of the uncertainty regarding the real cost of natural gas and oil. As these costs became more clear in the future, detailed consideration should be given to the economics of underground hydrogen storage.

HYDRIDE STORAGE

While the natural underground storage of hydrogen probably represents the best potential for the broad energy reserve base of the solar-hydrogen economy, the intermetallic storage of hydrogen (metal hydride storage) represents perhaps the best potential for storage at the point of use. Intermetallic storage differs fundamentally from techniques so far considered in that it does not involve hydrogen gas under pressure.

It has been known for several years that certain metals form stable compounds with hydrogen. Relatively recently some of these metals and their alloys have been discovered to form hydrogen compounds readily at close to normal ambient temperatures and low pressure. Some of these compounds store hydrogen in greater density than is possible with liquid hydrogen and readily release the stored hydrogen with the application of an external heat source. These metal-hydrides offer many potential advantages over conventional means of hydrogen storage. They can be conveniently and safely stored, they can be made to release hydrogen under controlled and safe conditions, and the heat release and absorption processes which are associated with the formation of hydrogen compound can be applied in heat exchange systems.

In order to understand the exciting potentials of the metal hydrides we will first examine the basic processes involved in the formation of metal-hydrogen compounds and the types of metals involved. We will then review the technology of the process application and the potential uses. Finally we will consider the research directions which are being pursued in the development of new technologies and application.

We have previously considered the atomic structure of hydrogen and the chemical property associated with that structure called hydrogen bonding. Hydrogen possesses one electron in its first orbital and this configuration results in its tendency to form compounds with most of the other elements. Some of the compounds which hydrogen forms have unique properties which result from patterns of electron sharing and residual nuclear charges. We noted that water is one of these compounds. Another group of compounds which have unique properties are the hydrogen-metal compounds called hydrides. Some of these compounds, while stable under most circumstances are quite reversible with the application of heat or pressure. It is this stability and reversibility that makes these compounds particularly interesting as hydrogen storage systems.

The hydrides can be classified as covalent hydrides, saline or ionic hydrides, and metalic hydrides. Most covalent hydrides have such low boiling points that they exist as gases at normal ambient temperatures. These hydrides are largely the transition metals.

The saline hydrides are primarily the alkali and alkaline earth metals. The crystal formations of these hydrides are made up of ions of the metal and hydrogen and use a good deal of energy in the process of compound formation and decomposition.

The metallic hydrides have metallic properties such as metallic coloring and conductivity, but they are usually very brittle. There are two models of metallic hydride formation under general consideration at the present time. In one case hydrogen is assumed to donate an electron and form an alloy with the metal. In the other model, which is similar to the saline hydrides, hydrogen is assumed to accept an electron and exist as hydride anions, while the metal forms cations. The former model is called protonis and is supported by the fact that most metal hydrides are metallic conductors. The latter model is supported by the semiconductor properties of the rare earth trihydrides and the large enthalpies of formation of the metal hydrides.*

Of the many possible metals and metal combinations for metal hydride storage of hydrogen two of the most interesting at present are the binary metal hydrides: lanthanum nickel (LaN_{15}) and iron titanium (FeTi). Both of these hydrides have high storability of hydrogen with rapid hydrogen charge and removal characteristics at low temperatures and pressures. Both of these materials give off heat in the hydrogen absorption process equal to about 10 percent of the heat of combustion of hydrogen. This heat must be removed in the absorption phase and supplied in the desorption phase.[12] The heat flow in the process complicates the use of these metals but it also can be an added benefit by acting as a heat pump in energy systems which require heat manipulation. The necessity to add heat in the desorption process is also an important safety feature. In neither material is there a large volume of hydrogen at high pressure. Additionally, if the container system is ruptured the outrush of hydrogen reduces the internal pressure which increases the heat required for desorption and thus stops the production of hydrogen.[13]

Another hydride which has attracted considerable attention as a possible hydrogen storage system is magnesium (Mg). This metal is attractive primarily because it is light. However, the speed of hydrogen absorption of Mg is so slow that some other metal, often nickel (Ni) must be added to improve it. The addition of another metal increases cost and decreases storage capacity. Additionally, magnesium requires a great deal more heat in the process than the two intermetallics previously discussed. Even taking these limitations into account, the lower weight of magnesium, its relatively lower price, and high storage capability make it an important candidate.[14]

Certain shared characteristics of all hydride storage systems dictate the broad parameter of the technology necessary for their application. First their absorption/desorption of hydrogen are dependent on

---

* For a more detailed explanation see Arnulf J. Maeland, "Survey of the Different Types of Hydrides" *Hydrides for Energy Storage, Proceedings of an International Symposium held in Geilo, Norway,* August 1977, (Oxford, G. B.: Pergamon Press, 1978), pp. 19–28.

both temperature and pressure. At atmospheric pressure MgH_2 requires 300°C for hydrogen release, $LaNH_3$ about 12°C and FeTiH about -10°C. If the temperature is increased hydrogen can be released at higher pressure which may be of value for some uses. Second, while hydrides can be repeatedly used to absorb and desorb hydrogen this repeated use decreases the granule size of the material so that after a time the material has the form of a very fine powder.

These first two characteristics influence the technology of hydride containment. Containers must be strong enough to sustain the necessary temperatures and pressures. They must also be designed to withstand the horizontal pressure which develops as the granules become smaller and settle to the bottom of containers.[15] As a result of their smallness some of these particles may leave the storage cylinder with the gas, and this tendency to escape must be guarded against in the design.

A third important characteristic of hydrides is their tendency to become poisoned by impurities which may be present in the gas as a result of the method of generation. This is not significant problem with hydrogen produced by electrolysis. However, it becomes a serious consideration for hydrogen produced by coal gasification. We noted earlier that a primary means of purifying hydrogen produced from coal was through the use of metal hydrides. Hydrides used for this purpose of those used for the storage of hydrogen gas containing impurities must be periodically purged of the impurities which they absorb with the gas.

The metal hydride storage of hydrogen is barely out of the theoretical stage, but it offers great promise for many applications. There appears to be three major potential uses for metal hydrides at the present time. One of these uses, the purification of hydrogen produced by coal gasification, has already been discussed. The two other major potential uses are for hydrogen storage in specialized transportation systems and in residential storage.

Transportation machinery such as forklifts and hauling equipment which must be used underground or in enclosed spaces are obvious candidates to be powered by hydrogen which is stored in hydrides. The safety consideration and the need for nontoxic exhausts would seem to overcome the disadvantage of the added weight of the metal hydrides.[16]

The use of metal-hydride storage systems for individual and mass road transportation systems has generally been discounted in the near future.* However this potential use should not be discounted out of hand. Town cars and small buses powered by electric motors, which are driven by electricity produced in fuel cells using hydride-stored hydrogen, would seem to offer a number of advantages beginning with the better relative efficiency of electric motors over internal combustion engines. This efficiency advantage would seem to compensate considerably for the cost of the extra weight of the storage system as long as extensive fuel range was not a major consideration.

The most intriguing possibility for the use of hydride storage is in conjunction with residential energy systems. In such systems concerns for safety, reliability and cost are paramount whereas weight and size are less important. A residential energy system that produces electricity by photovoltaics, uses this electricity as needed, and then stores the remainder in a metal hydride for future conversion to electricity in fuel cells has clear advantages. Such a system can be totally independent of outside sources, extremely safe and, after an initial cost of purchase and installation, virtually nonpolluting. In addition, the heat of hydride absorption and desorption can be used to run a heat pump which can act as an air conditioner or a furnace when supplemented with passive and active solar thermal architectural designs.**

LIQUID STORAGE

Liquified hydrogen has major long-term uses in the transportation systems of a mature solar-hydrogen economy. These uses are primarily in the space industry, in aircraft and in rail transportation systems. While gaseous hydrogen is the product which is used, the liquid form is the most practical means of storage and transportation due to weight and bulk considerations. We have already noted how hydrogen is used in each of these systems and their advantages. We

---

* C. E. Lundin in the *Solar/Hydrogen System Assessment* does not envision wide spread use during this century. See Escher Section IV.

** The integrated photovoltaic, electrolysis/fuel cell metal hydride storage system for residential, and use being developed by the Texas Instrument Corporation is a case in point. See Chapter 2.

are concerned here with the special problems associated with the liquification of hydrogen, its storage, transportation and safety requirements.

Liquid hydrogen is colorless and volatile. It is one-fourteenth as dense as water and boils at about 20.3°K. The low temperature necessary for maintaining hydrogen in a liquid state requires special materials and conditions of storage. Since liquid hydrogen is cool enough to condense all other gases but helium it is necessary to assure that no air enters the storage area. Such entry would tend to concentrate liquid oxygen and create an explosive hazard. This is a problem when venting the hydrogen gas which naturally boils off as a result of the volatility of the liquid.[17]

Liquid hydrogen is usually stored in a double thermos tank called a *Dewar.* These Dewars are made up of two high-vacuum tanks. Liquid hydrogen is stored in the innermost of the two. They are separated by liquid nitrogen which aids in keeping the temperature low.[18]

While the procedures necessary for the storage and handling of liquid hydrogen have been well developed by the space industry some improvements in efficiency can be expected as larger systems are employed in support of hydrogen-fueled aircraft. The largest storage systems developed to date are two spherical containers located at the Kennedy Space Center, each with a capacity of 850,000 gallons. It has been estimated that a storage capacity of 3,000,000 gallons may be needed at airports to supply hydrogen-fueled aircraft. Such storage need not be centralized, however, and there appear to be no technical reasons why such a system should not be possible.[19]

Liquification procedures for hydrogen are well established but there are some unique aspects of cryogenic hydrogen which must be taken into account. It will be remembered that the hydrogen molecule can exist in two energy forms. In the *ortho* form of the molecule both atoms of hydrogen spin parallel to each other; in the *para* form the spin is opposite. The ortho form is a higher energy state. Hydrogen at room temperature is about 75% ortho and 25% para in composition. The ratio approaches 100% para at absolute zero. The process of conversion from ortho to para forms is not as rapid as the temperature decreases in the usual gas expansion method of liquification. If the process of conversion is not aided by a catalyst or allowed to adjust more slowly to the temperature change, then the liquid hydrogen will have the same 75% ortho to 25% para ratio as atmospheric hydrogen. The excess ortho hydrogen will continue to give off energy in the liquid form until equilibrium is reached. In the process the liquid will boil away at an artificially high rate. It is necessary therefore to cool liquid hydrogen to allow for the absorption of the excess energy given off during the conversion from ortho to para forms. This extra energy amounts to about 2% of the theoretical minimum energy required for conversion to a liquid state.[20]

Another problem inherent in hydrogen liquification has to do with the gas expansion process of liquification. Gas liquification takes place when compressed gases below a certain temperature (the Joule/Thompson inversion temperature for the particular gas) are allowed to expand. When this expansion takes place the temperature drops enough to cool the gas to the point of liquification. Hydrogen has a very low Joule/Thompson inversion temperature. Compressed hydrogen must be cooled to below 202°K before its temperature will drop on expansion.[21]

Notwithstanding these two complications in the hydrogen liquification process, the technology is well developed and dependable. The energy and environmental benefits of the use of liquid hydrogen in aircraft and rail transportation would seem to outweigh the technical difficulties. One need only consider the complexities of refining crude petroleum to make gasoline to see that the hydrogen liquification problems are not overwhelming.

TRANSPORTATION OF HYDROGEN

Hydrogen is frequently transported by rail in a gas or liquid form. The technology for such transportation is well developed and the safety record is commendable. However, except for certain specialized situations which may require the transportation of liquid hydrogen, by far the most important potential for the transportation of hydrogen is transmission through pipes.

An extensive network of pipelines in the United States deliver natural gas, primarily methane, from gas fields to points of use. Because this network is in place and represents a great deal of investment, it has been frequently suggested that it be used as a distribution system for hydrogen. The model which we will consider is based in part on this option and on the development of three major distribution corridors out of the existing network. One distribution pattern, called the *Appalachian Corridor* in the model, is developed out of the existing network of trunk lines which begins in the Gulf Coast area of Texas and Louisiana and spreads northeast to Pennsylvania. This corridor would service, as it does now to a large degree, the eastern seaboard and the use points east of the Mississippi. Another distribution pattern, which the model labels the *Missouri Valley Corridor,* builds on those networks which begin in west Texas and Oklahoma and spread northeast through the Midwest. The other major pattern beginning in west Texas spreads west along the Mexican border and up the California coast. This is called the *California Corridor* in the model.

In addition to the savings in cost there are several advantages to using the existing network. Many of these advantages are not readily apparent unless solar energy and hydrogen are considered together as a single energy economy. One argument against using the existing network has been that the network is based on the existing natural gas fields and that there is no reason for constructing hydrogen production plants, particularly coal gasification plants, in these areas. This argument overlooks one of the main advantages of the solar-hydrogen economy—its flexibility. Large *photovoltaic farms* using amorphous silicon plates, like those developed at RCA labs, combined with hydrogen generation systems could be located in the sunny areas of central Texas close to the beginning of the natural gas network. In addition, ocean thermal systems could be located in the gulf waters. Electricity developed in these systems could be used to produce hydrogen in plants located right at the source of the trunk lines. Supplementing and supporting these generation plants could be coal-gasification-to-hydrogen plants located in Illinois midway in the Appalachian Corridor. Large wind-generating systems which produce electricity for hydrogen conversion could be located in eastern Kansas, northwest Missouri and parts of Nebraska. Hydrogen from these plants could enter the system at the center of the Midwest Corridor. The California Corridor could be supported by both photovoltaics and ocean-technology hydrogen-generating systems all the way up the southern coast. This national production and distribution system, when combined with the smaller-mix, centralized regional and community generating systems, and individual residential units, would form the foundation of a mature solar-hydrogen economy.

The use of the existing natural gas pipelines for the transmission and distribution of hydrogen has been the subject of research and controversy for some time. The controversy centered on the questions of whether or not the existing system is compatible with hydrogen and whether or not it can deliver enough hydrogen to the effective. The first problem is primarily a concern with hydrogen embrittlement, a problem which we have discussed in conjunction with hydrogen storage. Embrittlement is a problem when pure hydrogen is pumped at high pressure through steel pipes. It can be controlled by reducing working pressures. A recent study indicates that the fracture toughness of pipeline steel is reduced by about 60% in 1000 psi hydrogen. The same study concludes the hydrogen can be transported through the natural gas pipeline network if working pressures are reduced.[22] The introduction of impurities such as carbon monoxide with the hydrogen also effectively inhibits embrittlement. The amounts of these impurities could be so small that they would be acceptable for all industrial or energy uses.[23]

At least two hydrogen pipelines are in operation. A low-pressure, 130-mile pipeline in the Ruhr Valley owned by Chemische Werke Huls has been in operation since about 1938 with no major incidents.[24] Another pipeline has been transporting hydrogen from Bayport to Baytown, Texas for some time.[25] Hydrogen has also been transmitted through conventional natural gas pipelines quite successfully. Until the end of the World War II in 1945, approximately 72,000 miles of pipeline in the United States were dedicated to the transmission of "town gas" which was about 50% hydrogen.[26]

There seems to be general consensus among the experts in the field that hydrogen can be transmitted safely through existing natural gas pipelines if small amounts of impurities are added or if the pressure of transmission is reduced. Reducing the pressure however raises the second important problem with the transmission of hydrogen through pipes.

Hydrogen has about one-third of the energy of an equal volume of methane (natural gas) even though it has much more energy by weight. This is a result of hydrogen's low density. As a result of this poor energy-per-volume relationship, three times as much hydrogen gas must be transmitted to equal natural gas. There are two ways to overcome this relative energy deficiency. First, more hydrogen can be transmitted per unit of time. Hydrogen is much less viscous than natural gas; therefore it moves faster through the pipe at the same pressure. The greater speed of transmission offsets the volume/weight problem to some extent but hydrogen still delivers only about 0.7% of the energy of natural gas under mutually optimum conditions.[17] Second, the pressure of hydrogen could be increased, thus increasing the flow but we have seen that embrittlement is a function of pressure as well as hydrogen purity. In order to increase the pressure in the pipeline we would also have to increase the impurities.

It seems clear from recent studies that the natural gas pipeline network can be used to transport hydrogen, though some modification of the existing process would probably be necessary to accommodate increased flow. Hydrogen transmission costs would probably be at least twice as high as the transmission costs for natural gas. These costs of course do not take into account either the cost of the gas or the costs associated with its use.

In visualizing the potentials of a solar-hydrogen economy it is helpful to consider the generation, transmission, storage and end use of hydrogen in a whole system context. Approaching this system from an end use perspective it is clear that the whole system progresses over time from the use of the heat of combustion of hydrogen in its early stages to the use of the electrical energy of ionic transfer in fuel cells in the mature stages. One of the natural advantages of the solar-hydrogen economy is that it can employ the same systems of large-scale transmission and storage through the use of pipelines and underground reservoirs in all stages of development. These systems are largely already in existence and can be supplemented over time by smaller but compatible storage systems such as metal hydride for home, community and regional use.

The technological challenges of the storage and transportation of hydrogen are certainly present, and much more work needs to be done in this area. But much has already been accomplished. The feasibility of using at least some of the already existing natural gas pipelines for transmission seems to be indicated as does the possibility of large-scale underground storage. Safe metal-hydride storage is a reality, and the potential for the development of lightweight, safe, glass microballon storage seems good. Many other systems of storage will probably develop as the potential of the whole system is realized.

NOTES

1. A. W. Thompson. "Materials for Hydrogen Service" *Hydrogen: Its Technology and Implications,* eds. K. E. Cox and K. D. Williamson, Jr., (Cleveland: CRC Press, 1977), pp. 86–87.
2. Ibid., p. 113–114.
3. W. J. D. Escher, et al., *Solar-Hydrogen Systems Technologies,* DOE/JPL-955492, Vol. II, Part II, (U. S. Department of Energy, 1980), pp. IV-29.
4. Ibid., pp. IV 25–26.
5. S. L. Robinson and J. J. Iannucci, *Technologies and Economics of Small-Scale Hydrogen Storage* (Sandia Laboratories: U. S. Department of Energy, December 1979), p. 10.
6. Robinson, pp. 12–13.
7. Escher, pp. IV–80.
8. For a much more detailed coverage of the geological requirements for underground storage see M. R. Tek, "Underground Storage of Hydrogen" *Solar/Hydrogen Systems Assessment,* eds. W. J. D. Escher et al., DOE/JPL-955492, Vol. 2, Part 2, (U. S. Department of Energy, 1980).
9. P. O. Carden and L. Patterson. "*Physical and Energy Aspects of Underground Hydrogen Storage,*" *Hydrogen Energy Progress, the Proceedings of the 3rd World Hydrogen Energy Conference,* Tokyo, June 1980, (NY: Pergamon Press, 1981), pp. 797–807.
10. Tek, pp. IV 102–103.
11. Tek, pp. IV-84.
12. R. L. Cohen and J. H. Wernick. "Hydrogen Storage Materials: Properties and Possibilities", *Science,* Vol. 214, No. 4525, December 1981, p. 1082.
13. *Ibid.*
14. Escher pp. IV–69.
15. Cohen and Wernick, p. 1085.
16. Escher, pp. IV–71.
17. Escher, pp. IV 37–38.
18. F. J. Edeskuty and K. D. Williamson. "*Liquid Hydrogen Storage and Transmission,*" *Hydrogen: Its Technology and Implications,* eds. K. E. Cox and K. D. Williamson, Jr., (Cleveland: CRC Press, 1977), pp. 54–55.
19. Escher, pp. VI 41–42.
20. Edeskuty, pp. 52–53.
21. Escher, pp. IV–37.
22. W. R. Hoover et al., *Hydrogen Compatibility of Structural Materials for Energy Storage and Transmission,* Materials Development Division 8312, (Sandia Laboratories, Livermore, Annual Report, 1978–79).

23. G. G. Luth, "Transmission of Gaseous Hydrogen," *Hydrogen: Its Technology and Implications,* eds. K. E. Cox and K. D. Williamson, Jr., (Cleveland: CRC Press, 1977), p. 8.
24. Ibid.
25. Luth, p. 8.
26. F. J. Edeskuty, ed. *Critical Review and Assessment of Problems in Hydrogen Energy Delivery Systems, Initial Report,* Los Alamos Scientific Laboratory, University of California, Los Alamos, New Mexico, August 1978 for U. S. Department of Energy, LA-7405-PR.
27. *Ibid.*

REFERENCES

Carden, P. O. and L. Patterson. Physical and Energy Aspects of Underground Hydrogen Storage. *Hydrogen Energy Progress, the Proceedings of the 3rd World Hydrogen Energy Conference.* Tokyo, Japan: June 1980. NY: Pergamon Press, 1981.

Cohen, R. L. and J. H. Wernick. "Hydrogen Storage Materials: Properties Possibilities." *Science.* Vol. 214. No. 4525. December, 1981.

Edeskuty, F. J., ed. *Critical Review and Assessment of Problems in Hydrogen Delivery Systems, Initial Report,* Los Alamos Scientific Laboratory, University of California, Los Alamos, New Mexico, August 1978 for U. S. Department of Energy; LA-7405-PR.

Edeskuty, F. J. and K. D. Williamson, Jr. "Liquid Hydrogen Storage and Transmission." *Hydrogen: Its Technology and Implications.* eds. K. E. Cox and K. D. Williamson, Jr. Cleveland: CRC Press, 1977.

Escher, W. J. D. et al. *Solar-Hydrogen Systems Technologies,* U. S. Department of Energy, DOE/JPL-955492, Vol. II, Part II, 1980.

Hoover, W. R. et al. *Hydrogen Compatibility of Structural Materials for Energy Storage and Transmission.* Materials Development Division 8312. Sandia Laboratories. Livermore. Annual Report 1978-9.

Luth, G. G. "Transmission of Gaseous Hydrogen." *Hydrogen: Its Technology and Implications.* eds. K. E. Cox and K. D. Williamson, Jr., Cleveland: CRC Press, 1977.

Robinson, S. L. and J. J. Iannucci. *Technologies and Economics of Small-Scale Hydrogen Storage.* Sandia Laboratories. U. S. Department of Energy, December 1979.

Tek, M. R., "Underground Storage of Hydrogen," *Solar-Hydrogen Systems Technologies.* DOE/JPL-955492. Vol. II. Part II.

Thompson, A. W. "Materials for Hydrogen Service," *Hydrogen: Its Technology and Implications.* eds. K. E. Cox and K. D. Williamson, J.R. Cleveland: CRC Press, 1977.

7. THE OUTLINE OF A MODEL

The first principle of thermodynamics holds that energy cannot be created or destroyed. According to this principle there is a fixed amount of energy in any *closed system,* and while this energy can change form, even to the point of becoming matter, the amount can never change.

The second principle, often called the second law, suggests that all energy processes progress from high-quality energy states to low-quality energy states. The quality of energy is a measure of the potential useful work of a state. Another way of conceiving the quality of energy is through the organizational behavior of molecules. In a high-quality energy state molecules possess organized behavior; in a low-quality state their behavior approaches randomness. Total randomness is called a state of *maximum entropy.* In this state all energy is in the form of unusable heat.

The concept of an *open system* does not necessarily challenge the principles of thermodynamics, but it does make their application significantly different. Open systems share similarities which distinguish them from closed systems in both structure and function. While open systems are also collections of dynamic processes and distinct from their environment, they have in addition a direct and mutual relationship with their environments which is not characteristic of closed systems. They have three distinguishing characteristics which result from this relationship with their environments. The first of these characteristics is *choice.* Open systems can reach a number of different states as a result, to a greater or lesser degree, of internal decision making. Another characteristic of open systems is *evolution.* Open systems can transit toward more complex states of organization. Evolution results in an entropy increase in their environments because energy has been taken in and used. Finally, open systems have *adaptability,* they can adjust to changes in their environments.

Open systems receive energy from their environment. They then incorporate this energy in some form of a control apparatus and act to produce an effect on their surroundings as a result of a combination of internal controls and energy received. An open system responds to the changes it produces in its environment through a process called *feedback* (Figure 7.1).

An energy economy is a system which functions to provide useful work in the human community. This work is usually in the form of heat, motion and light. In this book, we have termed the present global energy economy the *carbon-combustion economy* because its common denominator is carbon and its fundamental process is combustion.

For over a century the industrial world has existed in what might be termed a fossilized-hydrocarbon phase of the carbon-combustion economy. Because the fossilized-hydrocarbon phase is based on the use of large, easily accessible reservoirs of stored energy, we have not had to account for the laws of thermodynamics in the development of cost analysis for deliverable energy. In determining the cost of the energy available from oil and coal, the biological solar energy conversions are never counted because they have already been paid for over the millions of years it took to form these reserviors of coal, oil and natural gas. The basic costs have been those associated with the mining, processing, and delivery of the resource. Only recently have we begun to count some of the costs of the waste heat and chemical pollution resulting from the use of fossilized hydrocarbons. While we have treated the carbon-combustion economy, particularly the

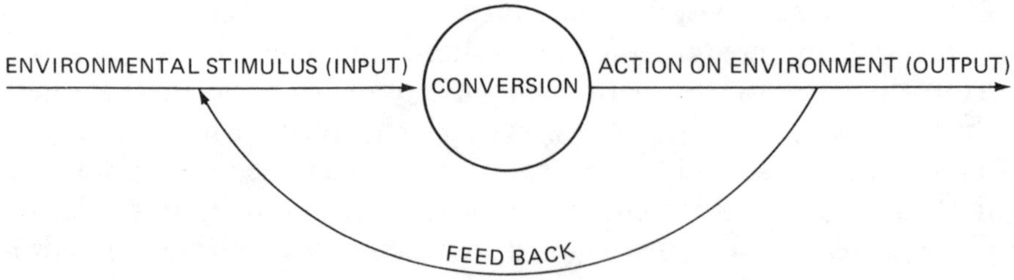

Figure 7.1 The concept of an open system

fossilized-hydrocarbon phase, as an open system with a free external energy source, it is in reality a closed system in which no new hydrocarbon fuels are accumulated.

The carbon-combustion energy economy has no effective feedback mechanism to relate energy production with resource availability. In addition, the system has little sensitivity to the state or condition of the planet. It not only consumes atmospheric oxygen, but it increases the entropy of the biosphere with waste heat. In short, the present human energy system, the carbon-combustion economy, can be thought of as a great random fire which consumes carbon and oxygen in addition to energy and creates waste heat and chemicals.

If the planet is viewed as an open system, it becomes apparent that a number of phenomena can be employed to produce usable energy. The model outline which follows is a true open energy system. This system uses solar energy as its primary source. Its control apparatus is a range of conversion technologies which we have examined and which are designed to provide great flexibility and potential for change. The feedback process in this system can rapidly react to the changing energy demands of human society. Most importantly, this system, in its mature stage, produces very little waste heat because there need be little or no combustion on the earth. The mature system functions almost entirely on photoelectrical, photobiotic, electrochemical, and mechanical conversion processes. Solar energy is either used when it is produced or it is stored as hydrogen for future use.

In energy terms this system is potentially less costly than the carbon-combustion economy because the solar-hydrogen conversion processes are specifically designed for the production of energy and not for the maintenance of complex life functions as are the solar biological conversion processes of photosynthesis upon which the carbon-combustion economy depends. While it takes energy to make hydrogen, this energy will increasingly be supplied by a source external to the planet—the sun. As photovoltaics and other solar conversion devices become more efficient and their construction less expensive, the energy costs of generating hydrogen by solar means is borne more by the solar system and less by this planet.

As the monetary cost of mining and processing hydrocarbon fuels increases, as dependency on foreign sources increases, and as the costs of pollution become more apparent, the developing technologies of the solar-hydrogen economy will naturally become less costly in monetary terms. However, this natural process of market adjustment may not happen rapidly enough to avoid some of the potentially disastrous consequences of remaining in a carbon-combustion economy. The delicate balance of world peace is constantly threatened by the potential conflict over Middle Eastern oil resources, and the environmental deterioration caused by the waste products of burning hydrocarbons proceeds at a chemical and not an economical pace. The model presented here is an example of one way the transition to a solar-hydrogen energy economy might be conscientiously planned and staged in two human generations. A brief look at some energy use patterns will provide some background for this model.

ENERGY USE AND SUPPLY: 1985–2080

Energy use predictions are unreliable. Data upon which predictions are based are scanty and often hard to obtain because much information is proprietary and guarded. Further, much information is technically complex and subject to various interpretations.

Another reason predictions are so unreliable is that they are necessarily forecasts of the efforts of future human beings and future human institutions engaged in the creative work of technology building. It is literally impossible to predict the social and economic forces which will drive the processes of technology building. Finally, energy predictions are unreliable because they almost always reflect the desires of the predictors. They are, at best, visions of the future based on the desires, dreams, values, and sometimes the ethics of the predictors. The solar-hydrogen model which is considered here is not a prediction. It is an outline of an alternative energy future based on technological, social, and economic possibilities. The likelihood of those possibilities is not speculated. Nevertheless, the model takes into account various energy projections in order to describe possible energy production and use systems.

A common unit used to describe present and projected energy use on a national scale is the *quad*. A quad is one quadrillion (10^{15}) *British thermal units (Btu),* and a Btu is the amount of energy necessary to raise the temperature of one pound of water by one degree Fahrenheit. Another common unit is the *joule;* 10^{18} joules are roughly equal to a quad. It is estimated that the United States consumed between 78 and 79 quads of energy in 1980. This consumption has been categorized in a number of different ways. Usually, most consumption categories include residential, commercial, and transportation uses. Future demand predictions are based on a great range of variables, but all of these variables are related, in one way or another, to projections of human population and projections of energy use patterns. The range of interpretations of these two major categories of variables is wide indeed. The most expansive interpretations, those which assume the greatest growth in the human population and the greatest increase in intensity and range of energy use, envision a greatly increased energy consumption over the next several decades. The most conservative interpretation predicts a much smaller growth in the human population and a greatly decreased per capita energy consumption. Between these two perspectives lies a wide variety of interpretations.

A review of the range of these projections gives us an understanding of the difference in views of energy futures. Recently the Union of Concerned Scientists compared energy projections for the year 2000 from a number of sources. These projections range from a high of 125 quads to a low of 80 quads. An even more recent study projects an energy consumption of 33 Q in the year 2000.[1] At this point in time probably all these projections are possible.

Since energy consumption has risen in the past, it may be assumed that it will continue to rise in the future unless additional influences intervene either to reduce human population growth significantly or to change the consumption patterns of society. Those investigators who predict an expansive energy future, necessarily project an ever-increasing supply of energy. Usually these predictions depend, in the short run, on an increase in the supply of fossil fuels and in the long run on nuclear fission and fusion.

Many people hold a more conservative position and suggest that supplies of oil will be limited by both the economics of extraction and world politics in the near future and that nuclear power is too dangerous and too expensive to develop for long-term use. These energy conservatives usually base their projections on one or more influences intervening in current energy consumption patterns. The three most commonly assumed influences are decreasing population growth, changes in energy use patterns which result in a reduction of per capita energy consumption, and the introduction of energy technologies based on renewable resources. Usually these intervening influences are considered to require some stimulation by government policy.

The model which we will consider is in a sense both energy conservative and expansive. Its focus is on the development of an alternative system for the production, distribution, and use of energy. The system proposed would require certain changes in energy use patterns but not necessarily a per capita reduction in energy consumption once the system has matured. However, in the interim staging period it would require a low energy use projection scenario. For this reason we will use a modified version of the energy supply projections developed in a recent study completed by the Friends of the Earth (FOE) and published in 1979. This study projects that the total primary energy use in the year 2050 will be roughly 47% of the 1975 use (about 33 quads) and is the lowest estimate so far offered.[2] Our model proposes the gradual decrease in energy produced by fossil fuels so that by the year T + 45 energy production from oil, coal, and natural gas is reduced to 64% of the total with a further decrease to 42% by T + 65. Conversely, the percentage of energy produced by solar energy is projected to increase to about 31% of the total by T + 45 and to increase to about 50% by T + 65. The model we will consider assumes a continued increase in energy production by solar means other than biological to almost 100% by the year T + 90.

THE SOLAR-HYDROGEN SYSTEM: OUTLINE OF A MODEL

It is ludicrous to assume that the energy system of any of the industrial societies of the world is a loosely organized mass of individual entrepreneurs and business people carving a livelihood through competition with each other in a fickle market of enlightened consumers. Energy systems in industrial societies are complex and highly controlled economies which have traditionally involved substantial planning.[3] Any changes proposed for such systems must account for the real nature of these systems including their control and planning mechanisms and the possible failure of these systems.

There are many alternative energy futures which might involve solar energy and hydrogen. We will examine a model system which involves maximum use of technological potentials within the outer boundaries of current social and economic feasibility. The model we will consider will undoubtedly strain the limits of popular acceptance and is offered less as a prediction than as a description of what might be if the United States exercised its best technical options for an energy future based on the solar-hydrogen system.

Many analysts now see some sort of solar-hydrogen system as part of a viable energy future. Other persons however remain skeptical about the possibility or advisability of completing a transition from a carbon-combustion to a solar-hydrogen system within the next century; nonetheless, most investigators are aware of the need to reduce dependence on the carbon atom. Some analysts, such as Wolf Hafele, director of the International Institute for Applied Systems Analysis, Luxemburg, Vienna, believe that a full transition from fossil to nonfossil resources will have to be completed before 2100 because resource constraints will become intense between 2030 and 2080.[4] With these thoughts in mind we will consider a model of a solar-hydrogen economy which reaches a mature stage about a century after a decision to implement.

There are four purposes for considering this model. First, it is offered to describe a systematic use pattern for solar-hydrogen technology. The model will relate these technologies to each other and to energy use demands of human society within the life spans of most individuals born during the next three decades. A second, and related, reason for the model is to indicate areas where research and development are needed in order to make a solar-hydrogen economy a viable energy future.

A third purpose for the model is to provide a basis for examining the problems and possibilities associated with energy policy development. Finally, the model will help us visualize some potentials for an energy future which might not be otherwise considered. Once again, the model should not be viewed as a policy proposal. It is instead a way of examining possibilities.

The model solar-hydrogen economy will describe three distinct time periods. The starting point of the model is represented by a concept called *time of commitment* (T). In real planning terms T would be a calendar date. The three time periods span about one century: (1) the primary implementation staging period, T to T + 15 years; (2) the middle period, T + 15 to T + 45 years; and (3) the staging of a mature solar-hydrogen economy, T + 45 to T + 95 years. Each of these periods will be briefly described with regard to the general physical location of energy generation and distribution systems and the percentage and projected total energy produced by each of five basic conversion systems (Figure 7.2).

The basic solar conversion processes used in this model have been selected because they are generally regarded as likely possibilities for implementation over the next several years. They are also compatible with the production of hydrogen over a range of techniques. These five conversion processes include biological, wind, water, heat, and photoelectric systems.

Biological systems for the conversion of sunlight into energy are the most complex of the processes considered. They also include the most familiar energy systems currently in use—fossil fuels. The use of coal, oil, and natural gas as fuels in the solar-hydrogen economy is a necessity during the first two time periods. However, these fuels are gradually phased out so that by the last stage, the mature solar-hydrogen economy, they are no longer used. Other biological conversion systems which emulate the processes inherent in photosynthesis will be phased into the economy as they become technologically available.

Wind power is projected as an important and continuing conversion process throughout the whole model. It is expected to increase in all stages.

Water power is also considered to be fundamental throughout the model. While all other conversion processes with the exception of coal are envisioned as both centralized and decentralized sources, water power is primarily a centralized system. Both ocean thermal technology and kinetic technology (waves and tides) are considered for implementation.

Photoelectric systems are perhaps the most flexible of the conversion processes projected in the model. These systems are projected to continue to develop over the next several decades. The cost over time of energy produced by photoelectric, solar thermal, and wind and water electrolysis systems has been projected in a study completed for

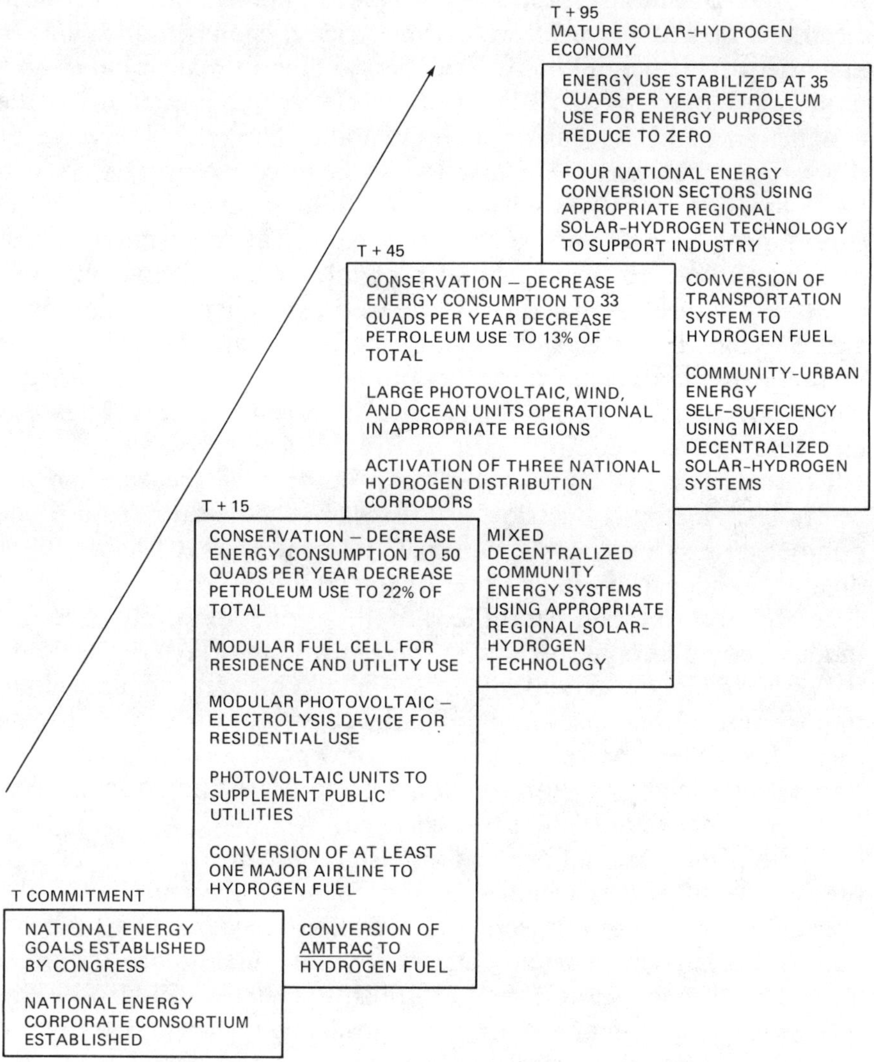

Figure 7.2 Outline of a model plan for staging a solar-hydrogen economy

the U. S. Department of Energy by the Jet Propulsion Laboratory. It is estimated that large photovoltaic electrolytic units, using advanced electrolyzer technology and photovoltaic equipment which produce electricity at $.02 per peak watt of electricity, could produce energy at a price of $25 per million Btu of energy. This estimate is based on a model developed by the Electric Power Research Institute for the production of hydrogen for the commodity market.[5] The value of this estimate lies not in its projection of the price energy but rather in its recognition of the viability of such a system. Even if the price of energy produced by such a system remained at the estimated best price of $25 per 10^6 Btu it would probably be competitive with oil, coal, and natural gas by the late 1990s on the basis of dollar price alone. What the estimate does not account for are the costs of the *market externalities* such as air pollution and national security problems associated with energy dependence on foreign sources. These externalities are inherent in the use of fossil fuels.

Photoelectric/hydrogen systems can be both centralized and decentralized energy conversion processes. In the decentralized mode photoelectric systems could be staged into the market for use in individual residences in most areas of the country.

In addition to this decentralized mode of operation, photovoltaic modules could be concentrated in *solar farms* in the area of the country with the highest annual rate of sunlight. This area, which encompasses some of the most arid land in the country, is in a strip running through southern Arizona, New Mexico, and Texas. It is a natural photoelectric energy sector and could produce vast amounts of electrical energy for conversion to hydrogen.

Finally, solar thermal conversion systems are projected as increasing sources of energy in all three time periods. Decentralized solar thermal systems, both passive and active, are recognized as increasingly important for providing energy for individual homes and other structures of habitation. Small centralizing systems will also be employed in the same manner. Larger centralizing systems are projected to provide increasing quantities of electricity and heat later in the model. In the mature solar-hydrogen economy, hydrogen will also be produced through the process of the *thermal dissociation* of water.

The model projects a continuous increase in the use of hydrogen as a means of storing and transporting energy. The system for this storage and transportation generally reflects an expansion and modification of the system already in use for the national distribution of natural gas. The model also projects large hydrogen production systems located in the Gulf and central Texas close to the most concentrated network of existing natural gas pipelines. Hydrogen could be shipped through this network along three major corridors which are presently used for the distribution of natural gas. One of these corridors, which we will call the *Appalachian Corridor,* lies along the interior side of the Appalachian Mountains and would service the East Coast. Another, the *Missouri Valley Corridor,* passes through the Midwest to the Great Lakes region. Finally, the *California Corridor* would serve the coastal area west of the Rocky Mountains.

The model system which we will examine is focused on the development of a mature solar-hydrogen economy within a century after commitment to the system. An important step in this development is the implementation of a solar-hydrogen production, distribution, and use system on a national scale in the United States by T + 45 years. This second-stage system will be based on four major solar energy conversion elements: photovoltaic devices, wind generators, hydropower systems, and solar-thermal systems. These elements would produce heat and electrical energy for onsite use. They would also be centralized for the production of hydrogen in modular electrolytic elements for transportion via pipe lines to underground storage areas along use corridors. Reconversion to electricity would take place through the use of modular fuel cells which would be used alone or grouped together to produce large amounts of electricity. Free hydrogen in liquid form would be used to power aircraft and to power fuel cells to operate rail transportation systems.

Inherent in the second stage would be a national research effort to develop advanced photovoltaic systems, photobiotic systems, and other more refined technologies for direct hydrogen production. While coal is still an important source of hydrogen in this stage, the emphasis is on the transition from fossil fuel to other forms of solar energy. The third stage of the model projects the complete national disassociation from the carbon-combustion energy economy through the implementation of a fully mature solar-hydrogen economy.

The beginning of the model period, between T and T + 15, is a key period because it is during this time that the near-term staging of technology implementation must take place.

The model thus has three periods which will be considered individually. Each of these periods will be viewed with regard to energy conversion, energy storage, energy transportation, and energy use.

Staging Primary Implementation, The T to T + 15 Year Period

The first-period stage is characterized by a public decision to implement an energy future founded on solar energy. This decision is identified as the *time of commitment* (T), and it is accompanied by a decision to commit public funds to the process of implementation through the establishment of a *National Energy Corporate Consortium* (NECC) with the power to actively compete with multinational energy corporations on a global scale. This national consortium would be commissioned by the Congress to invest public funds in basic and applied research which would be undertaken by a research organization of the consortium in cooperation with aspects of higher education on university campuses. The consortium would be directed to operate in cooperation with a restructured, small-business administration to encourage, and if necessary, capitalize new energy corporations with the specific purpose of implementing new technology in the solar-hydrogen area.

A series of national goals for the year T + 15 would be established by the Congress, and the Consortium would be commissioned to monitor their achievement. Among these goals would be:

1. The decrease in national energy consumption to 50 quads per year through conservation;
2. The decrease in the consumption of petroleum, both domestic and foreign, to 22% of the total energy use;
3. The development of the following energy use profile:

A. Natural gas	15.2%
B. Coal combustion	32.0%
C. Coal gasification to hydrogen	20.0%
D. Solar thermal	2.0%

E. Water power	7.6%
F. Wind power	.8%
G. Solar electricity	.4%
H. Oil	22.0%

4. The conversion of 5% of the energy generated by natural gas, coal, and water power to hydrogen for storage and transportation;
5. The development of a commercially available, modular fuel cell for use in residence and utilities;
6. The conversion of the entire Amtrak system to hydrogen fuel;
7. The development of commercially available modular photovoltaic units for use in public utilities and private residences;
8. The development of commercially available modular reversible electro-chemical conversion devices (ECD) which could be used to generate electricity for immediate use, convert it to hydrogen for storage, and convert hydrogen to electricity for residential use;
9. The conversion of one major airline to hydrogen fuel.

The emphasis of the first period is on the development of public acceptance of the solar-hydrogen option. Use patterns are gradually encouraged in the most readily implemented areas such as mass transit and individual residence use. Strong concern would be focused on the development of use patterns which are both safe and convenient. This stage would not see a great increase in hydrogen for home use but would rather focus on broader energy use patterns which could be more closely monitored while safety systems were refined. Conversely, individual home use of photovoltaic devices would be encouraged by every possible means.

Most of the hydrogen produced in this stage would be by coal gasification to hydrogen, first as a mixture of carbon monoxide and hydrogen (water gas) and later as pure hydrogen.

In the last decade of this stage a series of demonstration technologies would be introduced to the market, including an individual transportation system which used a fuel cell to convert free hydrogen to electricity for power, and an individual home energy conversion system consisting of an electrolysis/fuel cell combined with a hydride or glass-bead hydrogen storage system.

Staging the Middle Period, T + 15 to T + 45 Years

The middle period is the stage of implementation which envisions the general social acceptance of the solar-hydrogen economy. Whereas the first stage was characterized by the gradual development and introduction of solar-hydrogen use patterns through mass transportation, the second stage introduces hydrogen to general use through the conversion of at least 20% of energy production to hydrogen.

In T + 15 years the Congress establishes a new set of goals and restructures the National Energy Corporation Consortium. The consortium focus is shifted toward the implementation of new technology on a large scale, and is given two major objectives to accomplish by T + 45 years:

1. The development of a series of large demonstration projects including:
   A. Large photovoltaic, electrolytic units of at least one gigawatt capacity appropriately located in the Southwest and the activation of the Appalachian use corridor on a regular basis;
   B. Wind generation electrolysis energy systems of at least 1 gigawatt capacity located in midwestern states with connection to both the Missouri Valley use corridor and the California use corridor;
   C. Single- and multiple-stage, solar-thermal, hydrogen-producing generators located in the east central area of Texas.
2. The establishment of fifty mixed centralized community energy systems employing solar-hydrogen technologies appropriate to the specific region.

Energy use goals for this time period are established as follows:

1. A decrease through conservation and efficiency measures of national energy consumption to 33 quads per year;
2. The development of the following energy use profile:

A. Natural gas	3%
B. Coal gasification to hydrogen	52%
C. Solar thermal	14%

D.	Water power	11%
E.	Wind power	4%
F.	Solar Electric	3%
G.	Oil	13%

During this period most of the hydrogen produced continues to come from coal but electrolysis units are introduced on a much larger scale. The use of petroleum as an energy source is reduced to 13% of the total, and the reduction (almost 10%) is absorbed by conservation and solar conversion processes. This shift, while relatively small, is critical because it prepares for a much greater future implementation of solar-hydrogen technologies.

Staging A Mature Solar-Hydrogen Economy, T + 45 to T + 95 Years

The transition to a mature solar-hydrogen economy involves some radical shifts in energy use behavior. The foundation for these shifts has been established in the previous two stages. During this stage overall energy consumption is increased two quads to a total of 35 quads, but the use of petroleum and coal is reduced to zero.

Energy use goals for this period are established as follows:

A.	Solar thermal	25%
B.	Water power	30%
C.	Wind power	20%
D.	Solar electricity	25%

During the third stage a restructured National Energy Consortium, now composed of all major energy corporations licensed to operate in the United States, is commissioned by the Congress to implement the final stages of the solar-hydrogen economy based on the results of the demonstration projects and the market penetration efforts of previous stages.

Four major energy conversion systems are established in the United States during this stage. The *South Central Photoelectric Sector* located in New Mexico and Texas is established as a photoelectric/electrolysis product area. Hydrogen produced in this area is shipped via pipelines along all three use corridors. The use corridors are upgraded during this stage and constellations of reconversion centers, composed of banks of modular fuel cells, are established near population centers on the use corridors.

The *Mid Central Wind Power System* is established in Utah and Wyoming for the centralized production of electrical energy, part of which is used to produce hydrogen. *Water power energy production* is on a mixed centralized basis with four technologies implemented to take advantage of varying local conditions. Traditional hydropower, particularly low head, is used in areas of the Northwest and Northeast. Wave, tidal and OTEC systems are established near coastal population centers as supplemental mixed centralized power systems. *Solar thermal energy* as a decentralized energy source remains important, but increased emphasis is placed on thermal water splitting so that 60% of the solar thermal energy supply is used to produce hydrogen by T + 95 years.

The transition from a carbon-combustion economy to a solar-hydrogen economy is an awesome concept. It would involve considerable social, political and economic reorientation. In the next section we will consider some problems and strategies for technology development during the critical first stage of the model.

NOTES

1. John S. Steinhart, et. al. *Pathway to Energy Sufficiency, the 2050 Study,* San Francisco, Friends of the Earth, 1979.
2. Ibid.
3. John M. Blair, *The Control of Oil,* (New York: Randon House, 1977).
4. Wolf Hafele, "A Global and Long Range Picture of Energy Developments," *Science,* Vol. 209, July 4, 1980, pp. 174–209.
5. R. W. Foster et al., *Solar-Hydrogen Systems for the 1985–2000 Time Frame.* U.S. Department of Energy, June 1980.

REFERENCES

Axell, Elyse. "Chevy Purrs on 200 Proof." *Soft Energy Notes.* June/July 1981. pp. 94–95.

Behling, David J. *Analysis of Past and Expected Future Trends in U. S. Energy Consumption, 1947–2000.* Upton, New York: Brookhaven National Laboratory, 1977.

Bockris, J. O'M. *Energy Options, Real Economics and the Solar-Hydrogen System.* New York: John Wiley and Sons, 1980.

Chalmers, Bruce. "The Photovoltaic Generation of Electricity." *Scientific American.* Vol. 228. No. 1. June 1973, pp. 34–43.

Clark, Wilson. *The Energy and Defense Project. Energy, Vulnerability and War.* Washington, D.C.: Federal Emergency Management Act, 1982.

Department of Energy. *Energy Supply and Demand in the Midterm: 1985, 1990, and 1995.* Washington: U. S. Government Printing Office, 1979.

Department of Energy. *Inventory of Power Plants in the United States 1980 Annual.* Washington: U. S. Government Printing Office, 1980.

Department of Energy. *National Energy Plan II.* Washington: U. S. Government Printing Office, 1979.

Department of Energy. *Steam-Electric Plant Construction Costs and Annual Production Expenses - 1978.* Washington: U. S. Government Printing Office, 1979.

Department of Energy. *The Feasibility of Financing Domestic Energy Development During 1978–1984.* Washington: U. S. Government Printing Office, 1979.

Department of Energy. *The International Energy Evaluation System.* Washington: U. S. Government Printing Office, 1979.

Edmunds, Stahrl and John Letey. *Environmental Administration.* New York: McGraw-Hill, 1973.

Escher, W. J. D. et al. *Solar-Hydrogen Systems Technologies.* Vol. II. Part II. U. S. Department of Energy, June 2, 1980.

Finch, Douglas L. "The Japanese Photovoltaic Threat." *Solar Age.* February 1981, pp. 34–37.

Foster, R. W. et al. *Solar/Hydrogen Systems for the 1985–2000 Time Frame.* U. S. Department of Energy. June 2, 1980.

Gregory, Derek P. "The Hydrogen Economy." *Scientific American.* Vol. 228. No. 1. June 1973, pp. 13–21.

Hafele, Wolf. "A Global and Long-Range Picture of Energy Developments." *Science.* Vol. 209. July 4, 1980, pp. 174–209.

"Hydrogen Could Supply 10% of Ontario's Current Energy Imports by 2000." *Synfuels.* December 18, 1981, p. 7.

"Hydrogen Update, Microsphere Storage Investigated." *Hydrogen Progress Magazine.* Winter 1979–1980.

Gilmore, Robert W. "Power to the People: The promise of Decentralized Solar." *Technology Review.* August/September 1981, pp. 37–44.

Ku, W. S. et al. *Hydrogen Supplementation System Evaluation Study.* Upton, New York: Brookhaven National Laboratory, 1977.

Library of Congress. *Societal Aspects of Hydrogen Energy Systems.* Washington: U. S. Government Printing Office, 1974.

Lovins, Amory. *Soft Energy Paths, Toward a Durable Peace,* New York, Harper and Row, 1971.

"Luminescent Collectors as Photovoltaic Concentrators," *In Review: A SERI Research Update.* September/October 1981, Vol. 111/5.

Motor Vehicle Manufacturers Association of the United States, Inc. *World Motor*

Vehicle Data 1981 Edition. Detroit, Michigan: Public Affairs Division of the Motor Vehicle Manufacturers Association of the United States, Inc., 1981.

Muller, Lee. "Hubbub at Texas Instruments." *Solar Age*. December 1974, pp. 16–17.

Okagi, Alan and Jim Benson, eds., *County Energy Plan Guide Book Creating a Renewable Energy Future*. Fairfox, VA: Institute for Ecological Policies, 1979.

Patterson, Gordon N. *The Future Role of Hydrogen Fuel in an Electrical Society*. U. S. Department of Commerce, October 1974.

Salzano, F. and C. Brown, eds. *Hydrogen Energy Assessment*. Upton, New York: Brookhaven National Laboratory, 1977.

Salzano, F. J. et al., *Hydrogen Technology: An Overview*. Upton, New York: Brookhaven National Laboratory, 1979.

Smith, Jeffrey L. "Photovoltaics." *Science*. Vol 212. June 26, 1981, pp. 1472–1478.

The Energy and Defense Project. *Disposed, Decentralized, and Renewable Energy Sources: Alternatives to National Vulnerability and War*. Washington, D. C.: Federal Emergency Management Agency, 1980.

The Report of the Demand and Conservation Panel to the Committee on Nuclear and Alternative Energy Systems National Research Council. *Alternative Energy Demand Futures to 2010*. Washington, D. C.: National Academy of Sciences, 1979.

The Report of the Supply and Delivery Panel to the Committee on Nuclear and Alternative Energy Systems National Research Council. *U. S. Energy Supply Prospects to 2010*. Washington, D.C.: National Academy of Sciences, 1979.

8. TECHNOLOGY DEVELOPMENT

"... the discovery with which we are dealing involves forces of nature too dangerous to fit into any of our usual concepts." (The Congressional Record of 1875, concerning the discovery of gasoline). *

The rapid expansion of the electronics industry and the concurrent growth of powerful information processing systems, advances in biological engineering and medicine, robotization and computerization of production technologies—all of these and many more developments promise a new age of technology[1] Much of the impact of this new age will be in the area of energy systems; however, there are such large social, political and economic investments in the technologies of the carbon-combustion energy economy that safe transition into a new energy era will require extraordinary planning and management. Another crucial issue will be the choice of alternative energy futures.

The dangers inherent in the carbon-combustion economy have been described and need not be detailed here. It will suffice to point out again that continued dependence on the Middle East for petroleum places our national security in jeopardy and that the use of petroleum in energy systems constitutes a continuing environmental hazard. Delay in making a transition to a new energy economy is dangerous.[2]

The tempo of technology development is a function of multiple forces, some of which are not accidental. Smart business does not wait for a product market to develop out of the mist of supply and demand. Rather it creates both the supply and the demand out of

---

* Quoted by Anthony M. Moos in *Fuel Cells,* Vol. II, ed. G. J. Young, (New York: Reinhold Publishing Corporation, 1963).

capital and market potential. But this sort of market manipulation has its limits. Market potential is a calculus of forces which loses definition the farther ahead it is projected. Therefore, decisions to try to market new products are usually based on short-term considerations which are designed to optimize the safety of the investment. While risk and venture activity are not uncommon, it is not often that a major corporation will risk a secure profit-making activity through massive investment in a new and possibly competing technology. It is just this sort of investment however that will be necessary in order to make a rapid and safe transition to a new energy economy.

Corporate conservatism with regard to new technology is entirely rational. Private corporations have a single obligation—to produce a profit for their stockholders. They have no obligation to innovate new, socially beneficial technologies.

The range of new technologies which make up the solar-hydrogen economy may or may not lend themselves to short-term profit making. They do however promise social benefits which provide a compelling argument for the involvement of the public policy-making process in their development well beyond what has already been supplied. Three primary potential benefits in the solar-hydrogen economy support the need to make this transition as soon as possible. First, the solar-hydrogen economy offers the possibility of a safe, nonpolluting, inexhaustible energy supply. This energy supply is founded on an open-system thermodynamic process and can be designed to meet the needs of almost any degree of centralization. It has the potential of providing individual residences with independent energy as well as supplying the centralized needs of industry. The range of choices available within the system need be limited only by the demands and preferences of the users. Secondly, the solar-hydrogen economy is not dependent on foreign sources in any way. The independence it can provide would greatly enhance national security and reduce the risk of global conflict. Thirdly, and of immediate impact, the development and production of the technologies inherent in the solar-hydrogen economy would provide new jobs, new professions, and new futures for people and businesses throughout the country.

The possibility that these benefits may not be an inducement to the private sector of the economy has led us to consider some strategies for technology development which include the active involvement of the public policy-making process. This in no way indicates a lack of regard for the importance of the private sector. Private corporations are a fundamentally independent agent in the development of new technologies. They do however operate under the constraints of a decision-making process which limits their organizational ability to account for the larger public good in decisions concerning the development and implementation of new technology.

In order to make the jump into a new energy technology era, two efforts would seem to be required. The first is to develop and exercise an effective capacity of the public sector to make public policy on the basis of long-range public need and potential rather than short-term profit. It should be noted that the intervention of the public sector has historically been involved in all major technological innovation. A brief review of the petroleum industry's relationship with both state and local government, particularly in the mid-to-late-nineteenth century, would suffice to make this point.

The second required effort is to plan and structure the implementation of new technology in order that a minimum of social dissonance is created in the process. The history of the oil industry makes this last point particularly important. Oil is wild commodity. Beyond the fact that we know that it is a limited resource, we do not know how much remains or exactly where it is.

The oil market has been characterized by violent price fluctuations arising out of shortage and glut. Stabilization of the market, while maximizing profit, has been the underlying objective of the petroleum corporations since the development of the Texas oil fields at the turn of the century. The accomplishment of this objective throughout most of the last century has been a remarkable tour de force of cooperation. A landmark of this cooperation was the Achnacarry agreements in Scotland in 1928 which established the foundation for the "system" of market control within broad limits (the "as is agreement"). The inherent wildness of the oil market was somewhat under control, even throughout World War II, but with the breakdown of nineteenth-century colonialism and the development of OPEC, the world market controls began to fail.

At the present time, the fundamental interests of the players in the world petroleum market are split at least four different ways. OPEC itself is split into two camps with differing interests. The superrich countries such as Saudi Arabia share little in common with poor countries such as Nigeria. In the industrialized world there are two major players—the Soviet Union and the great oil companies of the West. The existence of such divergent interests in a potentially wild market, which has such influence over other markets such as natural gas, has created the potential for some violent activity in the energy economy during the next few years.[3] We might expect the price of oil to drop drastically in the next few years as cooperation in OPEC deteriorates. This drop in price would logically be followed by a severe shortage of petroleum and vastly increased prices. These market fluctuations will have little to do with the real availability of the resource or with its effects on health and environment.

If a decision is made to plan a transition to a new energy economy in the next decade, the first stages of the transition will probably take place during some of the fluctuations of the oil market; this timing will make the staging both more difficult and more necessary. The expression of effective public policy making will be particularly important if the energy market itself is unstable due to undependable supplies of oil and natural gas.

There are many forms which increased public sector involvment might take. One form might be an expanded planning role for the Congress and the formation of a National Energy Corporate Consortium such as that suggested in the model. Whatever form the public policy-making process might assume it would be necessary to operate according to a broad but clear strategy for technology development in the first stage of transition to a new energy economy. A careful review of the elements in a solar-hydrogen economy would indicate that such a strategy would have to address at least the following five broad requirements:

BROAD REQUIREMENTS FOR DEVELOPING A SOLAR-HYDROGEN ECONOMY

1. Planning

The development of the carbon-combustion economy was largely the result of the pursuit of the easiest path. Trees were burnable and available, therefore they were used for fuel. When forests were denuded, coal became the easy fuel, and the use of oil was a natural next step. Oil was convenient, cheap, and available. We have therefore never thought of energy as a resource which requires much planning. The fundamental premise of this book however is that the days of the easy availability of hydro-carbon fuels are nearing an end.

A solar-hydrogen energy economy involves a number of new technologies. Many of these new technologies are already developed, but their implementation will require substantial adjustments in the energy delivery and consumption patterns currently in use. Many other technologies are in early or beginning stages of development and will require staged research efforts. Developmental planning capabilities must be structured with a long enough duration in order that sequential progress in product development is effectively pursued while at the same time the flexibility to integrate new discovery is preserved.

Such a capability represents an extremely delicate balance of forces and probably will require the cooperative coordination of aspects of the private sector of the economy as well as public institutions. The problem is not unlike those confronted in the pursuit of national defense. The example of the development of the Polaris submarine is a case in point. Something like Polaris was determined to be an essential component of a national defense system. Yet the technology for the construction of a nuclear submarine capable of launching missiles from under the sea was not developed. The Polaris project was an example of effective long-range planning and implementation. The techniques used in this project have application for energy systems.

Before the Polaris project could begin there had to be general agreement in Congress and the Defense Department that such a project deserved extraordinary support. It was determined that the national defense need for such a system was so great that the public interest would be served by an extraordinary, long-term development effort. An effective, if somewhat rigid, planning tool called Program Evaluation and Review Technique (PERT) was developed for use in the project and considerable personnel and institutional resources were devoted to an effort which was ultimately successful.

The transition to a new energy economy could be viewed in a similar fashion—as an activity essential to the national security and directly in the public interest. Once such a view is accepted, extraordinary measures could be undertaken to coordinate the activities of the public and private sector in the accomplishment of a series of national energy goals.

Such an activity need not undermine the democratic nature of United States society because it would proceed from a recognized public mandate and progress toward the accomplishment of a public need similar to national defense, police and fire protection, and other public goods. Such a process could indeed be a step toward greater public involvement in issues which vitally affect the lives of everyone, but which are now controlled by chance events and short-term interests. In addition, the process could be an open one so that discoveries and developments made in the pursuit of long-range objectives could find their way rapidly into the marketplace through small businesses devoted to single products and specialized technological systems.

No planning process can ever be designed to account for all variables or to assure that a particular specific end goal is reached over a long period of time. What long-range comprehensive planning can do is create a set of near-term conditions designed to encourage the development of a particular general system state in the future. The near-term conditions which would be pursued in the long-range planning process of the sample model are developed in the fifteen years represented as the first stage (T–T+15). These conditions are expressed as *six basic technology delivery systems* whose development is considered essential for the growth of the solar-hydrogen economy over the next several decades. Later we will discuss these six technology delivery systems and show how they relate to the primary energy conversion patterns.

2. Compatability

Any long-term basic change in the energy economy must be phased over a period of time so that the shock of the change does not violently disrupt political and economic systems. Such management of change is a great challenge, and its success is largely a measure of the effectiveness of the planning process. Of particular importance for a solar-hydrogen economy in this regard are the *basic technology delivery systems* which are proposed for development during the first stages of the process. These systems should be designed to accomplish the optimum technology transfer into the economy while maintaining an essential compatability with existing systems. Where possible, such delivery systems should tie into existing energy patterns so that there is a natural retooling of industrial processes. In some cases, such as the coal gasification for the production of hydrogen, these ties are fairly clear-cut. In other cases, such as the development of photovoltaics-electrolysis-fuel cell systems, the linkages are not as clear. Even with the entrance of the petroleum companies into the photovoltaic business in recent years, there would still appear to be some basic, potential corporate conflicts developing between the powerful petroleum corporations and the growing electronics companies which are on the leading edge of photovoltaics research. To the extent that this conflict encourages the petroleum corporations to undertake effective research and development of photovoltaic systems, rather than to create strategies to impede the development of new technology, the competition will be valuable. Much will depend on the ability of the petroleum corporations to make the adjustment to new market potentials. However, the conflict could disintegrate into fights over control of the new technologies which would both discourage the development of new technology and damage the health of the old.

The basic technology delivery systems proposed for the first stage of the model are designed to integrate into existing transportation, residential, and utility use systems. This integration would be structured to encourage the introduction of changes in the use systems over a period of time necessary for the optimum use of existing commodities, the retooling of production systems, and for the retraining of individuals employed in the affected industries.

3. Safety

When considering the safety of the systems which make up an energy economy we must take into account short- and long-range effects which are not readily evident on the surface. Environmental denigration and long-term cancer-causing agents like radiation and chemical waste must be part of such considerations, along with concern for the physical safety requirements inherent in the operation of the systems.

One of the most positive features of the solar-hydrogen economy is that the systems which would compose it are considerably less dangerous environmentally than those which make up either the carbon-combustion energy economy or the nuclear energy economy. The serious environmental and health problems associated with these latter two energy economies are among the more persuasive arguments for the developing a solar-hydrogen economy. Nevertheless, it is essential that safety be a central consideration in the overall strategy for staging a solar-hydrogen economy. Such a consideration is all the more important in view of the unfamiliarity of the public with the technologies which make up the solar-hydrogen economy and the residual of public opinion which associates hydrogen with either the Hindenburg disaster or the hydrogen bomb.

Safety controls need to be given major design consideration in two fundamental areas. In the production of system components, care must be exercised to assure that production processes are environmentally sound and that dangerous by-products are either not created or are safely recycled soon after production. The other area where safety must be a primary consideration is in the production, distribution, and storage of hydrogen. While in many ways hydrogen is much safer than gasoline, the procedures for handling gasoline have been developed enough over the years that the relatively safe transportation, storage, and distribution are accepted as routine. Great care must be exercised to assure that safety is the primary consideration in the design parameters of a practical hydrogen-handling system.

4. Cost

Since the basic elements composing the solar-hydrogen energy systems are composed of some of the most available resources—sand and water (in the case of water the resource is recycled)—with the exception of relatively small amounts of platinum and other metals, most of the expense of solar-hydrogen energy systems is in research, development, and production. These expenses tend to concentrate at the beginning of technology implementation and are therefore capital demanding. This demand for upfront capital necessitates in part the involvement of the public sector. A serious developmental problem which has probably retarded the introduction of photovoltaics on the market has been what might be called the *first-in–first-to-lose* syndrome. A general assumption in the industries has been that major technological breakthroughs regarding efficiency and cost of production were right around the corner. No company wanted to be the first to develop and widely market a photovoltaic system which would be seriously obsolete before it realized a payoff of the development and production cost. Additionally, as we have earlier noted, there is a reluctance to expend great investment on research and development of products for which there is not yet a market.

The position of the federal government with regard to the involvement of the public sector in the development of solar technology has fluctuated greatly over the last several decades. In the early seventies, with the arrival of the first oil crisis and the formation of the U.S. Energy Department, there was a modest investment of public monies to encourage solar technology. This investment has been responsible for many of the newly developing solar-electric and hydrogen-producing devices which we have discussed. In the early 1980s we witnessed a great decrease in federal support for solar energy. The expenditure for solar dropped from almost 600 million dollars in 1982 to a projected 73 million dollars in 1983. It should be noted that this reduction did not represent an overall decrease in federal expenditures for energy; on the contrary, the proposed 1983 budget for nuclear fission was $1.3 billion and at least $12 billion went to tax benefits for oil companies.[4]

In view of the relatively small federal assistance over the years, the growth of solar technology has been remarkable. However, in order to take advantage of these developments it will be essential for the public sector to assist in the development of the technology delivery systems and in their introduction to the market. These systems should be developed with as much modularization as possible in order to allow for the introduction of system component improvements without the need to replace the whole unit. In terms of ecoomic benefit, the public assistance devoted to the development of these solar-hydrogen technology delivery systems should be great indeed. Unlike nuclear energy, which is massive and centralized, many of the solar-hydrogen delivery systems would be small enough to be marketed through a number of retail companies. They would also be composed of components which could be produced by a wide variety of companies, both small and large. In short, these systems have a broad and promising market potential at a reasonable cost. Exercising this potential in the short-term is one of the most important aspects of a strategy for the development of the technological systems of the solar-hydrogen economy.

5. Information

It has generally been assumed among scientists who are involved with hydrogen energy that the memory of the Hindenburg disaster looms large in the public consciousness and that most people have an exaggerated view of the dangers of hydrogen use because of this memory. The problem of public opinion and public knowledge with regard to hydrogen may however be even greater than previously thought.

In the fall of 1982 we undertook a representative public opinion survey of the adult residents of the City of Springfield, Illinois. We asked a random sample of 375 people, "Have you ever heard of hydrogen being used as a fuel?" 229 people, or about 61% said that they had heard of hydrogen being used a fuel. These 229 people were then asked in what way they had heard of hydrogen being used as a fuel. 51% answered, "the hydrogen bomb." This was a remarkable response. First, it indicated that 39% of the people surveyed had never heard of hydrogen being used as a fuel; secondly, it indicated that a majority of those who said they had heard of hydrogen being used as a fuel (fully 31% of the whole sample) associated the only energy use of hydrogen with the bomb. This response is particularly important since no other energy use was even close to this percentage. We found that only 9.6% of the people in this probably representative Midwestern city associated hydrogen with a legitimate energy use: 0.3% associated it with fuel cells, 3.7% with use in aircraft, 0.5% with use in trains, 1.6% with use in automobiles, 0.3% with use in farm tractors, 0.8% with use as rocket fuel, and 2.4% with use in the space program.

As a result of additional questions, we found that more people (2.1%) thought hydrogen was obtained from mining than thought it was produced by electrolysis (1.3%). With regard to safety, 59% of the individuals surveyed felt that hydrogen was "very dangerous" while only 30.7% felt that gasoline was "very dangerous." Correspondingly, 49.9% felt that natural gas was "very dangerous" while only 46.1% felt that uranium was "very dangerous."[5]

The results of this survey indicate that at least in parts of Midwestern United States there are significant numbers of people who have little or no knowledge of the potential uses of hydrogen. It also indicates that there are large numbers of people who associate hydrogen only with "the bomb."

An interesting sidelight regarding this survey testifies to the power of the media. In the last stages of sampling, our surveyors found a remarkable increase in the number of respondents who had considerable knowledge of hydrogen and its uses. This increase was so significant that we began asking people where they had learned so much about hydrogen and its uses. The answer was quite surprising. A day or two earlier a popular national radio commentator (Paul Harvey) had devoted a major part of a program to the topic as a result of his discovery of some of the more recent technological developments.

While we were not able to measure the exact dimension of the impact of this broadcast on public opinion, its effect was pronounced enough to require termination of the survey at that point due to obvious skewing of the data results. Fortunately, we had already collected a large enough sample to yield a statistically valid representation of the population of the city.

Public opinion and knowledge are very important aspects of any marketing effort. The size and effectiveness of the advertising industry testifies to this fact. However, it must be remembered that public information regarding technology is not an activity which has traditionally been limited to the private sector. In the area of energy technology the federal government had an essential role in bringing electricity to most of the rural areas of the country through the Tennessee Valley Authority and the Rural Electrification Program. This effort involved a major and effective public information program. Such a program will be necessary for a timely transition to a solar-hydrogen economy.

BASIC TECHNOLOGY DELIVERY SYSTEMS

The five requirements for the development of a solar-hydrogen economy outline above are inherently dependent on the technologies themselves and on how these technologies are grouped together for implementation. We will next discuss six groupings of technologies which we have termed *basic technology delivery systems.* These systems are designed for development during the first stage of the model. They are composed of those technologies which at this point seem to offer the best opportunity for timely implementation. Each of these basic technology delivery systems is designed according to the conversion, storage, and transportation patterns discussed earlier. A review of these patterns will help clarify these designs.

We have discussed the technologies involved in seven primary energy conversion patterns leading to the production of hydrogen:

1. Coal gasification to hydrogen
2. Hydropower generation of electricity to electrolysis to hydrogen
3. Photovoltaic generation of electricity to electrolysis to hydrogen
4. Wave and tidal generation of electricity to electrolysis to hydrogen
5. Ocean thermal energy conversion to electricity to electrolysis to hydrogen
6. Solar thermal generation of electricity to electrolysis to hydrogen
7. Biotic and photobiotic generation of hydrogen

There were two primary technologies which we discussed for the conversion of hydrogen to useful energy:

1. Chemical conversion through rapid oxidation (combustion) to heat
2. Chemical conversion through ionic transfer to electricity (the fuel cell)

We examined three basic storage technologies:

1. Pressure storage
2. Liquid storage
3. Chemical storage, particularly in metal hydrides

Finally we discussed one basic technology for the transportation of hydrogen and that was by pipeline.

These conversion, storage, and transportation technologies can be combined in various ways to develop the basic technology delivery systems. The development of six such systems appears to be essential during the first stage of the transition to a solar-hydrogen energy economy. These six systems have application to different sectors of the economy and would have wide-ranging effect, in the short-term, on national energy use patterns. More important than this however they would provide the technological foundation for the development of the broad technological patterns necessary for the transition to a mature solar-hydrogen economy.

1. The Regional, Coal-Gasification, Hydrogen-Production and Distribution Systems

The technology for the development of these systems (Figure 8.1) needs only to be coordinated and implemented. Such systems could be established in regions where coal is plentiful, preferably near a

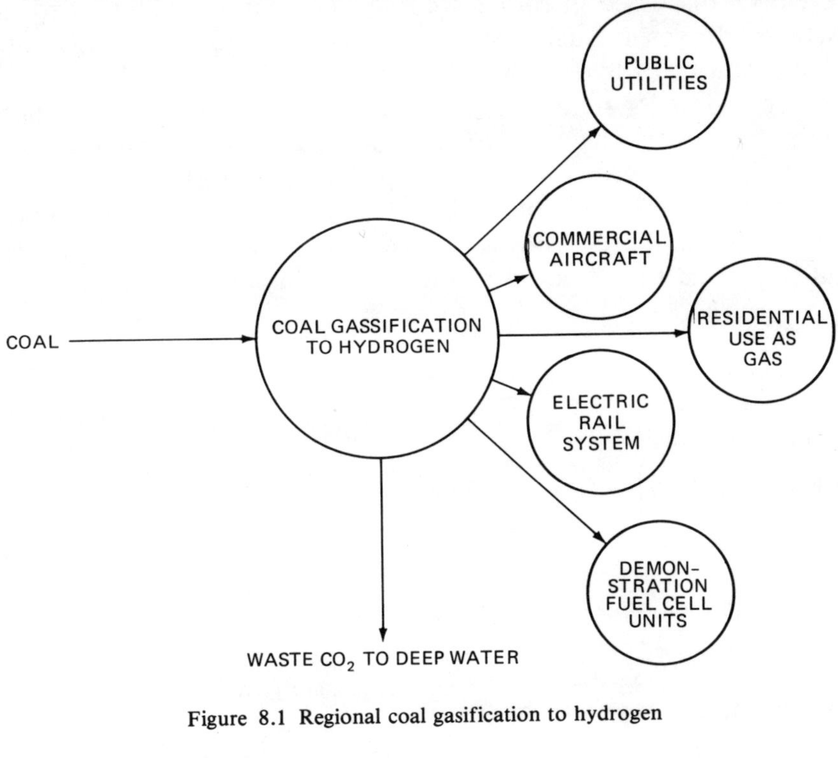

Figure 8.1 Regional coal gasification to hydrogen

large body of water where the waste carbon dioxide could be disposed of. Since almost any degree of hydrogen purity is possible in such systems as these, mixtures of hydrogen and carbon monoxide in about the proportions of "water gas" could be introduced into the natural gas distribution system immediately and used exactly the way natural gas is now used. As other technology delivery systems are introduced and the regional distribution systems are improved, the purity of hydrogen could be increased until all impurities were phased out. At that point, safe catalytic home heating systems and fuel cells could be used in regional residences.

However, the greatest benefit of this system would probably be its effect on the development of two other basic technology transfer systems. Both of these involve the transportation sector of the economy.

2. The Commercial Aircraft Network Using Liquid Hydrogen

The advantages of using liquid hydrogen to fuel commercial aircraft have been discussed in earlier sections. The major roadblock to the use of hydrogen in such a system has been the lack of a large and reliable source of hydrogen.

Once regional systems for coal gasification to hydrogen have been established, these regions can be connected by a commercial air transportation network fueled by liquid hydrogen (Figure 8.2). Hydrogen produced in regional plants would be piped to the regional airport, and at a processing area located nearby it would be purified and converted to the liquid form. The liquid would be pumped the

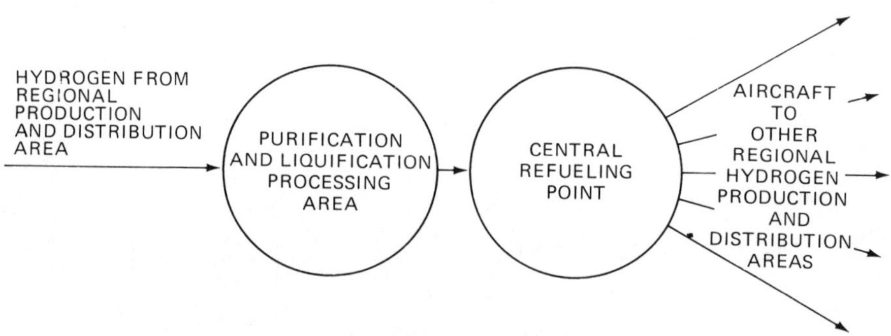

Figure 8.2 Commercial aircraft network using hydrogen fuel

short distance from the processing area to a central refueling point at the airport. All hydrogen-fueled aircraft would use this central refueling point where safety conditions could be carefully controlled.

As the number of regional coal gasification systems increased, the hydrogen-fueled aircraft network could be expanded. Such a network has particular applicability in the midwestern states where coal is plentiful and air traffic would justify the development of such a system.

A hydrogen aircraft network is not however the only transportation system which could develop as a result of a dependable supply of hydrogen. The other major transportation system would be a regional railway network, which could be both intra- and inter-urban. This system however will depend on another basic technology delivery system, a reliable and modular fuel cell.

3. The Modular Fuel Cell

As we have seen, fuel cell technology has advanced greatly in recent years. The development of solid polymer electrolytes have greatly improved both the durability and the size of such systems. The development of a compact fuel cell unit, which could be used alone for generating small amounts of electricity or combined with other units to generate large amounts, has clear advantages (Figure 8.3). These modular fuel cells could become integral parts of many technology patterns. Their use in the transportation sector of the economy and in public utilities is advantageous, particularly in the area of rail transportation.

4. The Hydrogen Electric Rail Transit System

Rail locomotives are now powered by electric motors. These motors receive their electricity from generators which are run by heat developed by burning diesel fuel. Electricity to power these motors could be developed in hydrogen fuel cells (Figure 8.4).

Hydrogen, which has been produced in regional coal gasification plants or by other means, could be piped to a processing area, purified, and converted into a liquid form. It could then be hauled in tank cars (as it routinely is now) and used as the fuel for banks of

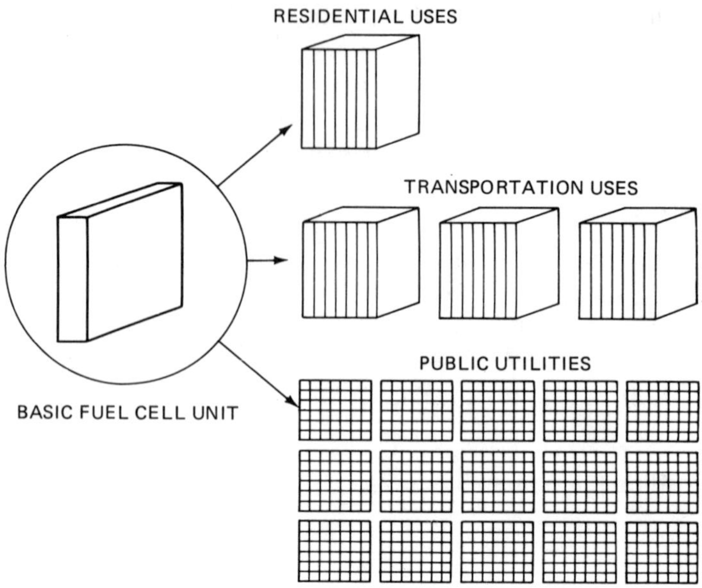

Figure 8.3 Modular fuel cell concept

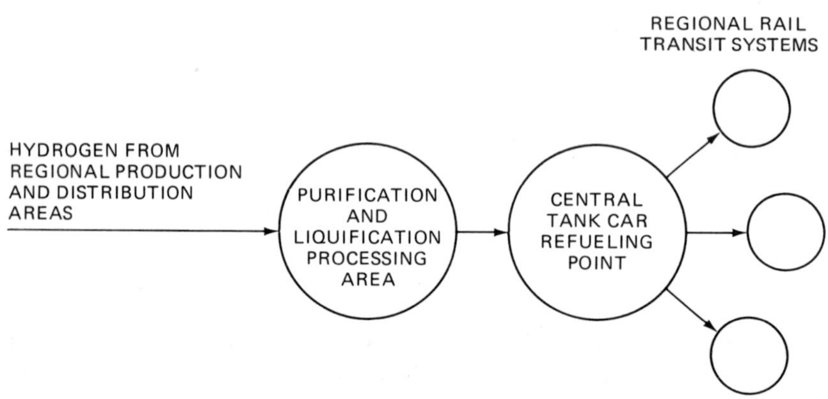

Figure 8.4 Hydrogen-electric rail transit system

modular fuel cell units to generate the power to run motors. Again, this system has all the advantages of hydrogen use and it is also compatible with major aspects of currently employed technology.

5. The Solar Energy System for Use in Public Utilities

The development of a system for the solar generation of electricity on a scale applicable to public utility use would be important for regional and small utilities. Such a system could be developed in the near future (Figure 8.5).

The heart of a large production photovoltaic unit is a photoelectric conversion device which can be cheaply constructed and which is efficient enough to generate adequate electrical current while not requiring an inordinate amount of surface area. We have briefly discussed recent breakthroughs by RCA Laboratories and others in the development of amorphous-silicon, photovoltaic generating systems with relatively high efficiency. These developments are significant enough to indicate that photovoltaic systems for use in utilities could be developed in the near future. These systems could be introduced on a small scale to help smooth out utility load patterns. In the long run they could be used exclusively to generate electricity for use in communities and small regions in which short transmission distance is required.

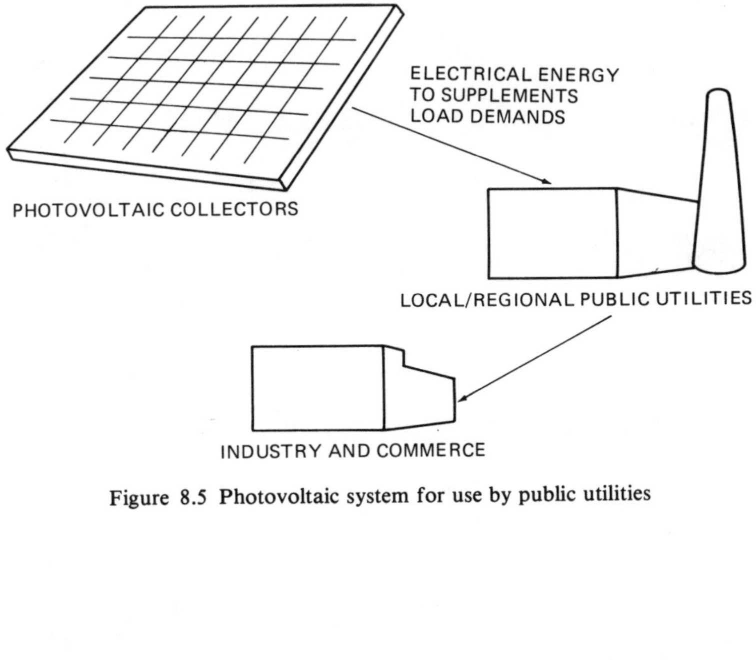

Figure 8.5 Photovoltaic system for use by public utilities

6. The Modularized Solar-Hydrogen Residential Unit

Perhaps no other technological possibility offers more hope for a sane new energy future than this one (Figure 8.6). Such a system would be composed of roof- or yard-mounted photovoltaic units with either integrated hydrogen production systems, such as those under development at Texas A & M University and Texas Instruments, or separate photovoltaics and hydrogen-generating systems. Hydrogen generated in such systems would be stored in stationary metal-hydride tanks. When needed, the hydrogen could be reconverted into electricity in fuel cells which could either be separate modular units or integrated electrochemical conversion devices (ECD) which both generate hydrogen and reconvert it to electricity. The heat exchange inherent in the metal hydride storage process could be used to power a heat pump to both cool and heat a residence.

These solar-hydrogen residential units could be available in the near future if the recent technological developments are capitalized and implemented. While such units may initially be luxury items for a small number of energy consumers, they could become popular products as construction costs decrease.

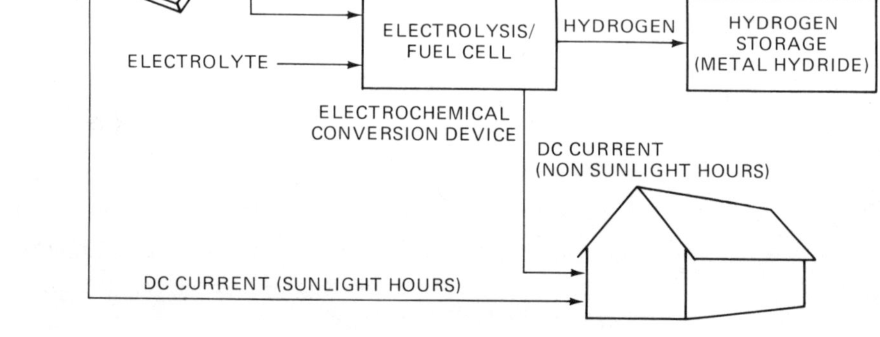

Figure 8.6 Modularized solar-hydrogen residential unit.

These six basic technology delivery systems have been selected because they seem to fulfill the requirements for an effective strategy in transition to a solar-hydrogen energy economy during the staging period of the model plan. They all fit the requirements of a coherent long-range energy plan, they are all more or less compatible with existing technological systems, while at the same time they provide a bridge to new and more wide-ranging technologies. Each of these systems either uses hydrogen distribution and storage processes, for which reliable safety patterns have been developed, or they use inherently safe systems such as hydride storage of hydrogen.

CHOOSING FUTURES

There is an economic theory called the *long wave cycle* which holds that the economy fluctuates over about a 50-year period. At the end of a period of capital expansion and economic growth this theory holds that there is about a decade of economic uncertainty followed by a decade of depression and economic collapse, during which most of the old capital investments and plants are worn out or destroyed. After this period of collapse there is a revival, during which there is great demand for new products. The forces that operate this cycle, it is maintained, are inherent in the economies of industrialized nations and have been operating since the turn of the nineteenth century. The data which support the long wave economic cycle indicate that there have been three major economic fluctuations since 1800 which have resulted in economic depression and the collapse of the capital base. Such depressions occurred in the 1830s, the 1890s and the 1930s. These data indicate that the industrial world may again be approaching a period of economic depression.[6]

A historical case can be made that during periods of economic depression there is much greater public sector involvement in the economy. This involvement has taken various form, depending on the nature and severity of the collapse, the international conditions which exist at the time, and the nature of the nation-state involved. Whatever the type of public sector involvement, it seems to influence the nature of the next stage of technological development and economic growth. Such growth may be similar to the railroad expansion in the United States in the latter mid-nineteenth century or the fantastic growth of new products after World War II. What also seems clear is that there has been little planning for these periods of economic depression or for the sorts of government involvement which are exercised in the process of economic rejuvenation and technological innovation. As a consequence, the involvement of government usually results from, and grows out of, the disappointment, confusion, and frustration of the public when unemployment rises and goods are scarce. The sort of government action which comes from this pressure is often a reflection of the very disappointment, confusion and frustration that caused it. The kinds of solutions which such governments tend to create are short-range, crises-oriented, and often violent. These solutions have been effective in the past, even though they have often created great suffering.

There has never before been the preponderance of destructive military power now available to the governments of industrialized nations. It is clearly in the interest of world survival to assure that the sorts of actions which governments take during periods of economic stress, or that they take to avert such stress, be based on a careful, balanced, and long-range view of the problems involved. These actions could be founded on the planned management of technological change so that violent fluctuations of the economic long wave are avoided or dampened by the introduction of technologies which go beyond those of the industrial era in providing continuity of employment and environmental security.

It would appear that there are two major alternative energy futures which lie beyond the carbon-combustion economy: the solar-hydrogen economy and the nuclear economy. Each of these alternative futures seems to be an available option at this critical juncture of human history, and each of them would require substantial governmental intervention for its realization. The nuclear option is a creature of the industrial era. Large and centralized, with inherent, incremental obsolesence, it is subject to all of the forces of the long wave cycle. It is an expensive option in energy, monetary and environmental terms, and it is supported by a powerful lobby and a large foundation of capital investment. But the public, which should be the most important force involved in the making of policies which determine its own future, is becoming less convinced of the value and the safety of the nuclear option.

The solar-hydrogen option is as yet little understood but it has advantages which are positive enough to demand consideration. It offers the possibility of a creative, coherent, and positive high technology compatible with human energy needs and planetary survival.

NOTES

1. For an interesting and insightful coverage of the development of the antecedents of this new technological age see Trevor I. Williams, ed., *A History of Technology,* Vol. VI, Parts 1 and II, (Oxford: Clarendon Press, 1978).
2. Two important works which deal with the dangers inherent in the carbon combustion economy are Wilson Clark, *The Energy and Defense Project Energy Vulnerability and War,* (Washington, DC: Federal Emergency Management Agency, 1982) and Daniel Yergen and Martin Hallenbrand, eds., *Global Insecurity,* (Boston: Houghton Mifflin Company, 1982).
3. For an interesting discussion of the nature of the control mechanism in the oil market see John Blair, *The Control of Oil,* New York, Randon House, 1976.
4. Paul Choity et. al., *In These Times,* November 3–9, 1982.
5. David C. Thomas, *Telephone Survey of 375 Residents of Springfield, Illinois,* September 1982. Phone numbers were selected at random and the sample was designed to be representative of the population as a whole.
6. The long wave cycle was described by Nikolai Kondratieff. In the late 1970s the existence of the cycle was supported by work done by Jay W. Forrester at MIT in an econometric model called the "System Dynamic National Model." Forrester indicated the correlation of these models in 1980. See Jay W. Forrester, "*Innovation and the Economic Long Wave,*" *Planning Review, the Journal of North American Society for Corporate Planning,* November 1980.

INDEX

Acid rain, 4
Amorphous silicon, 51
Appalachian Corridor, 150, 165
Atomic orbital, 22

Basic technology delivery system, 184
Becquerel, 18
Billings, 123
Biochemical conversions, 76
  bacteriorhodopsin (purple membrane), 85
  coal, 78
    anthracite, 78
    bituminous, 78
    entrained flow system, 82
    fixed and moving beds, 82
    lignitic, 78
  hydrocarbons: oil/coal, 45
  hydrogasification, 81
  hydrogenase, 84
  nitrogenase, 84
  photosynthesis, 54, 84
  pyrolysis, 79
  rhodospirillium rubrum, 84
  in vitro systems, 85
  in vivo systems, 84
Bockris, J. O'M., 5, 109
Bohr, Niels, 22
Boron, 47

California Corridor, 150, 165
Carbon, 3
Carbon combustion energy economy, 1
Carnot cycle, 112
Catalytic combustion, 120
Cecil, W., 122
CL-400, 131
Closed systems, 155
CO_2, 43
Czochralski process, 44, 53

Daimler-Benz, 124
Dendritic growth process, 53
Dewar, 148

Einstein, Albert, 20, 40
Electrochemical conversions, 88
  alkaline electrolyzer, 95
  anion, 90
  anode, 91
  bipolar systems, 95
  breakdown potential, 89
  cathode, 91
  cation, 90
  cell current, 94
  cell resistance, 94
  cell voltage, 93
  conduction band, 89
  current density, 93, 94
  deprotonation, 90
  electrical current, 24, 89
  Electricite de France, 95
  electrochemical conversion device (ECD), 106, 109
  electrolytes, 90
  electrolytic ocean mining, 97
  energy gap, 89
  fuel cell, 30, 92, 100
  fuel cell diaphragm, 93
  Gaz de France, 95
  ions, 90
  isoprene, 93
  microstorage, 140
  Nafion®, 92, 105
  Ohm's law, 90

Electrochemical conversion (*cont.*)
oxidation, 91
reduction, 91
regenerative fuel cell, 106
solid polymer electrolyte (SPE), 92, 95
standard equivalent conductance, 90
sulfonic acid group, 93
sulphonated polystyrene, 105
unipolar systems, 94
valence band, 89
voltage, 89
water reforming, 107
WE-3 (General Electric mixed oxide catalyst), 96
Energy, 1
energy medium, 2
energy source, 1
Erren, Rudolph, 123
Escher, E. W., 5

Feedback, 156
Fields, 41, 42
Field, static electric, 42
Flashback, 126
Fresnel lens, 49

Glass microballons, 10, 138
Glow discharge, 51
Greenhouse effect, 54

Hertz, Heinrich, 18
Hindenburg, 9
Hollow fibers, 140
Hydroelectric power, 62
Hydrogen
bonding, 28
embrittlement, 139, 151
hydrides, 10, 138, 143, 144
hydrogeneration, 142
hydronium ion, 90
hydroxl ion, 91
ions, 28, 29
ortho and para states, 27
protium, tritium, 27

Johnson, Clarence L. (Kelly), 131
Joule/Thompson inversion temperature, 149

Kinetic energy, 56, 57
King, R. O., 123
Kondratieff, Nikolai, 193

Lateral pulling, 53
Liquid Hydrogen Experimental Airline Project (LEAP), 132
LeMay, Curtis, 131
Lenard, 18
Lockheed U-2, 130
Long wave cycle, 191
Lovins, Amory, 113

Market externalities, 12, 164
Maximum entropy, 155
Mid Central Wind Power System, 170
Missouri Valley Corridor, 150, 165

National Energy Corporate Consortium, 168
Nitrogen oxides, 119

Ocean current energy, 68
Ocean thermal energy conversion (OTEC), 55, 61, 66
Ohm's law, 89
Open systems, 155
Operation Suntan, 131
Otto cycle, 121
Ozone layer, 4, 119

Peroxyacetyl nitrate (PAN), 119
Photoelectric effect, 40
Photon, 20
Photovoltaics, 6
cell barrier, 47
conduction band, 45
diffusion length, 40, 45
doping, 47
electron hole, 40, 45
heterojunction, 48
homojunction, 48
MIS junction, 48
phosphorous, 47
photovoltaic farms, 150
Schottky barrier, 48
SIS junction, 48
Planck, Max, 19
Polycrystal silicon, 45
Pre-ignition, 127
Project Bee, 130
Protonis, 144

Quantum mechanics, 22

Rae, R. S., 131
Rankine cycle, 66
Reservior pore volume, 141
REX engine, 131
Ricardo, 122
Rutherford, E., 21

Schrodinger, Erwin, 22
Second law of thermodynamics, 155
Silane (SiH_4), 51
Silverstein, Abe, 130
SiO_2, 43
Slater ducks, 63
SO_2, 4
Solar ponds, 68
Solar thermal processes, 5
  acceptance angle, 71
  central receiver systems, 69
  centrifugal separation, 87
  compound parabolic concentrators, 71
  effective aperture area, 71
  linear focusing distributed receiver, 70
  membrane diffusion, 87
  parabolic dishes, 70
  point focusing distributed receiver, 70
  shift conversion, 107
  tracking heliostats, 69
South Central Photoelectric Sector, 169
Stoichiometric ratio, 117, 127

Teams to Advance Research for Gas Energy Transformation (TARGET), 106
Tek, M. R., 143
Thomson, J. J., 20
Tidal motion, 65
Thermal gradients, 66
Trihydrides, 145

Water power, 61
Water power energy production, 170
Wave power, 63
Wind conversion, 55
Wind velocity, 57